THE LASER
EXPERIMENTER'S
HANDBOOK

For: Tanya Lynn, Monica, and Ricky

THE LASER EXPERIMENTER'S HANDBOOK

BY FRANK G. McALEESE

TAB BOOKS

BLUE RIDGE SUMMIT, PA. 17214

FIRST EDITION

FIRST PRINTING—AUGUST 1979
SECOND PRINTING—MARCH 1980

Copyright© 1979 by TAB BOOKS

Printed in the United States of America

Library of Congress Cataloging in Publication Data

McAleese, Frank G.
    The laser experimenter's handbook.

    Includes index.
    1. Lasers—Handbooks, manuals, etc. 2. Lasers—
Experiments. I. Title.
TA1683.M33   621.36'6   79-17465
ISBN 0-8306-9770-5
ISBN 0-8306-1123-1 pbk.

Cover photo courtesy of Bell Laboratories.

Contents

Introduction

Before launching into our studies of the laser and its predecessor the maser, and conducting an appropriate investigation of their respective *ticking mechanisms*, I'd like to spell out certain intended parameters as to just what educational ends this *Laser Experimenter's Handbook* may carry to the reader and experimenter.

I have, by necessity, drawn certain preliminary assumptions in respect to the reader's academic background in the general sciences. A science background is an imperative prerequisite for any explorer in this complex field of study.

Without a doubt, this particular subject field, which deals with the inseparable aspects of atomic nature, molecular construction and electromagnetic wave propagation, is perhaps one of the most complex and technically profound of all the branches of science and physics.

Although one does not need to hold a Masters Degree in Microwave Physics to be able to construct a first class working laser from blueprints, one must nevertheless have a nominal degree of technical conversance and practical ability to facilitate the selection and assembly of parts.

You will be able to intelligently construct a laser or two, or even a dozen at the completion of this *Laser Experimenter's Handbook*. And you will, I promise you, have acquired along the way, a working familiarity of underlying concepts and theories of the laser as its technical anatomy is explored. Perhaps the reader will also adopt an inspired reverence for the interrelated faces of science which give life to this marvel of our era.

Were I so commercially inclined, I could just offer the enclosed laser plans, along with a brief commentary on the construction of each, provide a parts and suppliers list and call it a deal, but I personally feel that we, both as reader and author, would be short-changing ourselves in the process if I were to limit my goals to that narrow involvement and impersonal relationship. It is my hope that this brief introduction to the world of advanced microwave applications will encourage the reader and experimenter to embrace a more comprehensive attitude toward the careers of science and to advance to possible graduate studies in the fields of laser and maser technology.

Anyone at all could pick up a set of skillfully drawn schematics or drawings and fabricate a simple laser to some degree; but how much more gratifying it would be for us to understand the scientific phenomena we are utilizing in the projects.

By benefitting ourselves with a parallel education of the whys and wherefores of our efforts, we could then enjoy the satisfaction and rewards that were once totally realized only by those dedicated scientists and physicists who founded our present day schools of science, such as Hertz, Maxwell, Faraday, Planck and Einstein. Were we to settle for less than the whole package of education along with the mechanical entertainment, we would certainly be sacrificing an intellectual experience.

Again, I am providing the reader and experimenter with only the absolute related requisites for pre-laser and maser physics in this book. A general background of basic science and physics is an assumed holding on the reader, for it far exceeds the scope and function of this study to provide the experimenter with a lengthy exposure to the general sciences. The experimenter should understand the theoretical spirit of the laser and maser, be able to construct from my plans a functionally operating laser or maser and through an educated grasp, the experimenter should have the confidence to modify or expand his or her projects to whatever end established, with competence and safety.

This author wishes to sincerely encourage the experimenter to reinforce his or her knowledge in the applied sciences of the laser, for the more we suffice ourselves with the exposures of the related laser and maser phenomena and potential, the closer the statement will reveal itself in describing the laser and maser as solutions in pursuit of problems.

The possibilities and known benefits of the laser and maser in the fields of industry, medicine, communications and weaponry are far too numerous to even begin to mention here; but be assured, that we now stand on the shoreline of the new era of electronic physics, and to more than any other single instrument, or man-made device, our future conquests will be owed in great part to the laser and maser.

A glossary has been included to acquaint the reader with the many new and peculiar technical and descriptive terms that are somewhat exclusively related to the field of light and electromagnetic interests.

And now, let's embark on perhaps what might be the most fascinating exploration of our academic lives—the study of matter, energy and atomic nature.

Frank G. McAleese

Acknowledgements

It is with gratitude that I grasp the opportunity of this moment of recognition to offer my unrequitable appreciation and thanks for the generous and concerted contributions and expertise given so freely, from cover to cover, of this, my first educationally oriented book.

Thanks To: Bell Laboratories of Murray Hill, New Jersey; The Hughes Aircraft and Laser Co. of Carlsbad California; The R.C.A. Solid State Division of Somerville, New Jersey; Metrologic Instruments of Bellmar, New Jersey; The Western Electric Co. of New York City; Central Scientific Co. of Chicago, Illinois and Spectra-Physics of Mountain View, California.

And for their personal allocation of time, technical collaboration and much needed materials, including equal measures of encouragement and inspiration, thanks to the following precious individuals: Inspector Joseph Czop and entire Czop Clan of Blue Bell, Pennsylvania; Dr. W. C. Woods Jr., Professor of Science, Glassboro State College, New Jersey; Ms. Donna Brown, Professor of English, Cumberland County College, New Jersey; Mr. Tony Curtis, Editor, Tab Books, Blue Ridge Summit, Pennsylvania and Mr. W. "Mule" Pierce, Technical Consultant, Elmer, New Jersey.

Lastly, but with paramount respect, to the Grand Master Architect of this Universe, for His marvelous and profound wonders, to which this awe-struck student shall forever be a servant.

Chapter 1
Matter, Energy and Atomic Nature
Atoms—Atomic Excitation

The basic principles by which lasers and masers draw their life functions will be highlighted in this chapter. The verifiable interrelationship of mass and energy and the proven and workable coexistence they maintain in their naturally occurring forms will be discussed.

Matter, or that which is *sensible*, by some detectable or measurable means, and *Energy*, which in most of its forms is more elusive in its detection, *are* one and the same.

Matter is a physical representation of energy, and conversely, energy may be defined as an end product of matter, or an extended result of the essence of matter.

If we were to completely reduce, or disintegrate a given mass of matter (since it cannot in any way be totally destroyed), we would have remaining in its place, an equal quantity of energy. Not weight, but quantity. The principle of the Atomic Bomb is a very good example of this fact, whereby a given quantity of mass is almost instantly converted into the energy it represents.

We have learned then, that matter is but one *face of energy*, and thus, energy is but one *face of matter*. From this we could conclude that everything in the universe is a product of matter and energy. The total *product* or *resultant* of all the mass and all of the instantaneous prevalent energy equals a mathematical *constant* at any one given time. Think of it, in the entire expanse of this universe, there is a *constant* which when applied mathematically, will account for every gramweight of matter, multiplied by all of the existing energy at any particular instant!

Dr. Albert Einstein, scientist and mathematician, gave us a formula for computing this relationship of the interconvertibility, or conservation of matter and energy. He stated, or rather equated that the *energy available (potential) is equal to it's mass (in grams) multiplied by the speed of light, squared.* Or algebraically written: $E = M \times C^2$ where the speed of light, in a vacuum $= 186,000$ miles per second, or $300,000,000$ meters per second.

Let's say we wanted to know just how much energy we could, under theoretical conditions, squeeze out of a kilogram of coal:

$$E = mc^2$$
$$E = (1Kg) \, (3 \times 10^8 \text{ m/s}) \, (3 \times 10^8 \text{ m/s})$$
$$E = 9 \times 10^{16} \text{ j or nt-m}$$

one kilowatt hour equals 3.6×10^6 J. (one joule equals one ampere of current flowing across the difference of one volt.)

$$E = \frac{9 \times 10^{16}}{3.6 \times 10^6} = 2.5 \times 10^{10} \text{ Kw hours.}$$

Or, 25 billion kilowatt hours.

Even at the low cost of one cent per kilowatt hour, (which was about the price when Edison was developing light bulb filaments) its dollar value equals $250,000,000.000—*two hundred and fifty million dollars*!

The preceeding illustration is somewhat realized in Nuclear Power Plants, but is, as I pointed out, only hypothetical in every day drama.

Having set forth the seemingly illustrative, or transitory state that all matter momentarily exists in, I think we are ready to explore the more physical composition of matter, and later, we'll have a look at matters' counterpart, energy.

ATOMS

Matter, be it any item, substance, compound or thing, is basically, and fundamentally built from, or made up of, *molecules*.

Molecules can be said to be further made up of, or built up from, *atoms*. And by zooming in with our imaginary microscope we will switch to the next higher power, and state that atoms are made up of still smaller particles which we will identify as *protons*, *neutrons* and a very small speck of a thing orbiting about these two particles, which is named an *electron*. Here, look in the microscope for yourself.

In Fig. 1-1A, there is a common drop of water, which we will refer to as H_2O. By increasing the magnification power of the microscope we will witness that this drop of H_2O is in reality, composed of three parts: a center part of an oxygen atom and two atoms of hydrogen, which are attached in a *Mickey Mouse ear fashion* (Fig. 1-1B). After moving the magnification power up one more notch, we observe that what we once thought were just three items are now something like 13 much smaller units (Fig. 1-1C). These turn out to be three nuclei (the nuclei of two hydrogen atoms and one oxygen nucleus) and 10 electrons (eight electrons belonging to the oxygen atom, and one electron each belonging to each hydrogen atom).

In Fig. 1-1D we isolate one of the hydrogen atoms and find it to be composed of a planetary arrangement of a nucleus with only one *nucleon*, in this case a proton and with one electron orbiting about its outer periphery. By raising our power up to the final notch we see the oxygen atom in all its splendor, having a nucleus of 16 members: eight protons and eight neutrons (Fig. 1-1E). Since we must have a matching electron for each and every proton, we count 8 electrons. All that out of one little drop of water!.

Figure 1-1E clearly illustrates the basic atomic nature and geometric pattern of all atoms, comprised of a nucleus with its nucleon members—the proton and neutron, and orbiting electrons about its outer envelope.

Without delving into chemistry more than needed, I will refresh the reader's memory by stating that the very outer-

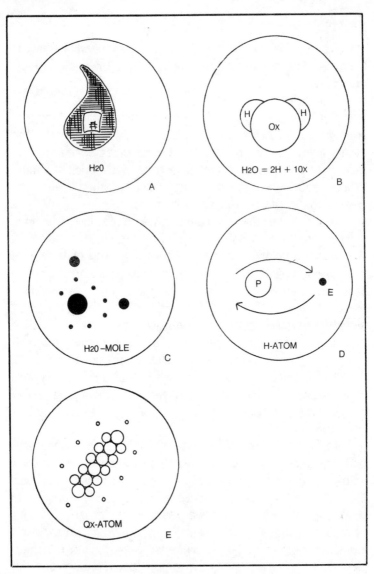

Fig. 1-1. A single drop of water as seen through a microscope.

most envelope of the electron flight plan or *orbital arrangement* serves as the combining, or *hooking* facility by which the individual atoms hook onto other atoms. They combine to form the thousands of different compounds and molecules of which matter and substance are made. The ability of any atom to unite, or join with another atom to form a molecule of a compound is known as its *valance*. This is nothing more than its number of electrons minus, or in excess of a total outer number of eight electrons, which appears to be a magic number in the atomic world. Thus, oxygen, which has a *valance* of minus two, will have, by virtue of its outer most six electrons, the ability to pick up an extra electron from a *donor-type* atom such as hydrogen. Since hydrogen only has but one electron to donate, it will require two hydrogen atoms to provide the two electrons that the oxygen atom needs to bring its outermost ring to a total of eight electrons.

We've had quite a microscopic introduction into matter, haven't we? But I've deliberately left out a very important part and for a good reason. I want to reintroduce you to the molecule in a different manner. The part I didn't mention was the portion called *energy*. It is energy which must be added or taken away from the atoms. The energy must be added externally or be released internally, depending on the process or compound in question.

But it's a basic enough process for our purposes, so we'll examine the situation by studying the proton in Fig. 1-2A. It is absolutely identical with any other proton that could be detached from any other atom of any other molecule, regardless of the individual type or character of matter it originally represented.

All protons, neutrons, and electrons are identical and equal in both qualitative and quantitative dimensions. The difference in the type or character of matter is governed by the number of individual particles and the geometrical manner in which they are arranged, on both the atomic and molecular

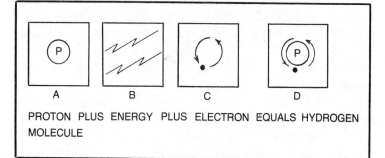

PROTON PLUS ENERGY PLUS ELECTRON EQUALS HYDROGEN MOLECULE

Fig. 1-2. The formation of a hydrogen molecule.

scale. A proton is a proton, a neutron is a neutron and an electron is an electron.

Having established that important issue, we will take the proton (Fig. 1-2A) and add a certain quantity of energy to it (Fig. 1-2B). Bring these two together with one captured electron (Fig. 1-2C) and you have a hydrogen molecule, complete with all necessary components—matter and energy (Fig. 1-2D).

The hydrogen atom was selected in this illustration because of its inherent simplicity, in that it has the least number of electrons and basic nuclear structure with just one proton.

It must be emphasized that all atoms of all elements are essentially similar in their multiple forms. They all have a nucleus of at least one proton, one or more neutrons and an equal number of electrons. The electrons are equal in number to the sum of their protons—from the lowly hydrogen atom, up to and including the uranium atom (235), which contains 92 protons, 143 neutrons and exactly 92 electrons. (One electron for each proton gives us the total atomic weight of 235).

All atoms, from the lightest to the heaviest have *mass* in common. Mass is expressed by its atomic weight, due to the respective quantity of protons and neutrons. The contributing weight of the electron is so negligible that it is disregarded. For the record, the electron mass is approximately only 1/1847 the weight of either a proton or a neutron. This means that it would take 1,847 electrons to equal the weight of one proton or neutron, so it's a justifiable neglect.

In addition to mass, an atom also contains energy. It contains two types of energy to be specific. One is the energy which is *housed in* the nucleus which holds the proton and neutron in their place, also known as *binding* or *packing* energy, which is what *nuclear energy* and *nuclear reactance* is all about. The lesser in value of potential is *electron energy*, which is easier to deal with and exploit than the atom's *nuclear energy*.

Nuclear energy deals with energy levels and values which require very elaborate devices such as Van de Graaff generators, cyclotrons and betatrons to either extract the atom's energy or rearrange its construction for other purposes.

We will confine our interests then to the atom's *electron energy*. This energy is the frequency by which our electron is vibrating at, or the distance the electron is from the nucleus in its orbiting flight, or in what energy level or *shell* our electron is residing. This affects its energy level by either its *vibrational frequencey* (Hz.) or its *shell occupation*. It requires more energy for the electron to move further away, or to occupy an outermost shell or level, than it requires for one closer to its influential exerting nucleus counterpart, or *home base*. When I say electron level, I'm referring to that level which the electron will occupy in its *natural state*. Another way of saying *natural state* is to say *normal excitation level*.

Normal excitation level, a new, but not really horrifying term for us to examine, is that state of being, or level, by which nature causes (with no outside interference) the molecule or atom to occur in. This state or conditional level of energy is affected by the molecule's temperature and the molecule's presence under any electronic or electromagnetic forces of the environment.

Thus, we could affect a change in, or upon the atom or molecule by changing either the temperature or electrical state. If we were to change either state or both, we would be changing the total value of the molecule's level or state of energy.

Since I mentioned temperature first, we'll deal with this affecting influence first. The temperature of a substance is a *direct indication of its energy* on an atomic and molecular level.

As we add heat to a substance, we raise its temperature and the speed or frequency of its vibrating molecules. The higher we raise the temperature of this substance, the more energy we impart to it. Eventually, before the substance melts or vaporizes, a point will be reached where, because of the unaccustomed new level of *thermal* excitement, the molecules become very *vibrant*. They commence to *glow*, having reached *incandescence*. This point of incandescence is caused by the changes in the atom's energy level, whereby the electrons are so vibrant and energetic that they fly further away from their normal orbital paths. They have absorbed the energy to do so from the external source of additive energy, which in this case, is the application of heat. The electron is now a further distance from the nucleus than where it would be under normal unexcited temperatures or circumstances.

In order for this electron to return back to its normal and closer radial distance or shell around the nucleus, it must give up that energy somehow. As at every shell or home orbit there is only a certain energy content which the residing electron of that orbit may possess. Any additional energy must be disposed of in some way. This is accomplished by ejecting or emitting the acquired extra energy. The electron should then return to its normal *home shell*, its closer radial distance of orbit.

The energy which the electron emits is in the form of a *packet* or small quantity of energy, which for obvious reasons we'll call a *quanta of energy*. This *quanta of energy* has been emitted on such a frequency as to place it in the visible light range of the *electromagnetic wave spectrum*. This simply means that these waves at this particular frequency can be visibly detected. Due to this *photo effect*, the *packet of energy* can be described as a *photon of energy*. That very superficially covers the *quanta theory of matter*.

We included earlier in this chapter the effect temperature had on matter in raising its energy level. In addition to temperature, or the effect of *thermal energy*, the energy level can be affected by means of applied *electrical energy*, or *light energy*. This leads us to believe that the energy level of an atom and its ultimate molecule can be affected by externally applied energy.

Before moving along to our next discovery, let us re-examine the facts covered thus far:

■ Atoms, which are the essential basic units of all molecules, are composed of two primary components—mass and energy. The mass they represent is in terms of weight, which is a direct and relevant "read out" of the sum total of protons and neutrons. Energy, that is electron energy, as opposed to nuclear energy, is the energy of the electron which acts as the medium or vehicle by which radiant energy will be exchanged or interconverted. This interconversion is by either internal or external sources, or by means of excitement.

■ The orbital shell or natural electron radius in its position from the nucleus acts an absorber or temporary storage facility for the absorbed energy. This energy may be externally applied to the atom which responds by increasing its diameter or dimension to accommodate any absorbed energy. This elastic-like force normally has a fixed limit or radius at a *ground level energy value*.

When this radius in stretched or expanded, the atom, by virtue of its orbiting electron, will return to its normal status or *ground level* state of existence by pulling the extended electron back to its formerly closer radius. For the electron to return to its original radius it must release its abnormally contained energy. It does this by emitting the energy in the form of a discrete packet or *quanta* of energy. Once it has discharged, or emitted this extra energy, the electron may return to an orbiting radius closer to the nucleus, thus returning the atom to its earlier unexcited state or to its *ground level of energy value*.

What happens to the quanta of energy emitted from the atom after the electron returns to its original energy shell and the atom has decayed from its excited state? Well, the Law of Conservation of Matter simply forbids its disappearance. The location of this photon will be discussed later.

In Fig. 1-3 there are three environments containing four atoms. The atoms are depicted as they normally exist in their normal, ground level state (Fig. 1-3A). Arrows of energy are directed or *applied* into this environment of ground level atoms. Figure 1-3B shows how the atoms appear after absorption by the arrows of energy which they were hit by. You can see how the individual atom appears after becoming excited by absorbing the applied energy (Fig. 1-3B at right). The electron has absorbed this energy and now orbits, or flies in an orbital shell. This shell is a radius of greater magnitude than what it normally would be in its unexcited, or ground level state. Thus, the electron and its orbital position have been expanded in order to accommodate this new level of energy.

The excited atom and its far-flying electron, as it now exists after releasing the momentarily absorbed energy, is seen in Fig. 1-3C. Notice the arrows of energy packets leaving the atoms. The individual electron has now returned to its closer or lower level orbit or shell, *only* after giving up or radiating the applied exciting energy which caused it to become excited in the first place.

ATOMIC EXCITEMENT

All atoms of matter normally exist at a certain energy level, or ground level. At least the greatest majority or *population* of the number of total atoms of a given mass of matter exists at this *population level*.

We can, by applying energy to this population, excite the total of atoms by way of *electron shell expansion*. They then go from their original ground level of existence to a new level of excitement, or energy content. The majority of atoms exist at a ground level of normally low excitement. The relatively small percentage of the population that exists at any higher level is due to the raising of the energy level by exciting the population. This causes the population of atoms to be *inverted* from their normal existence to a new level of excitement. Therefore the term *population inversion* tells us that we have *inverted* the *population* of atoms from the normally occurring small percentage of randomly excited atoms, to a new larger percentage of the population in the excited or *inverted* state.

We have witnessed that the atoms in Fig. 1-3C return to their former, more comfortable ground level after having once absorbed the applied energy, or after once having become excited. This return or reversion to ground level is known as *decaying*, which is a pretty good definition of the process. During the atom's *energy decay*, it must, as a requirement of returning to ground level, emit and dispose of the energy which was responsible for elevating the atom to its excitation level to begin with.

When an atom, or for that matter, the entire population of atoms, *revert* to their ground level status, they decay. In order for our population to decay, it must emit the stimulated energy which was responsible for its excitement. This emission and disposal of energy is accomplished in the process of decay by emitting the energy to its surrounding environment of neighboring atoms. Thus, one atom may be excited by applied

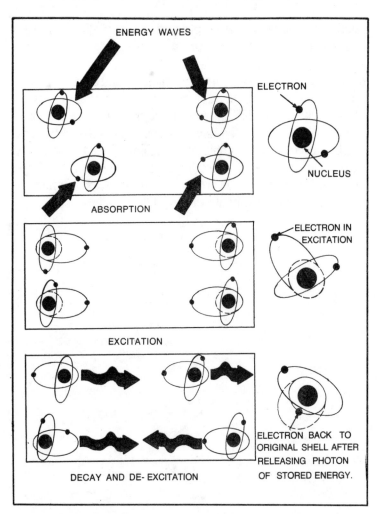

Fig. 1-3. Atoms and electrons go through absorption, excitation and decay in three separate environments.

energy, dispose of this energy and dump it off on a neighboring atom or molecule. This process of transference of energy is known as *reaction*. In this particular case, it would be called *population reaction*.

This should not be confused with *nuclear reaction*, because we are not dealing with energy on the nuclear level, but rather on the atomic level.

It is by the total comprehension of the terms in this chapter and their respective definitions that our ideas and the phenomena for which they represent will come alive. They will appear for us in our future appraisal and investigations into the ticking mechanism of the laser and maser. Without a solid familiarity of the underlying principles of matter, energy and atomic nature, we will be at an absolute and irreparable loss in dealing with the manifestations of the properties and characteristics for which these subjects represent. After all, light amplification by the Stimulated Emission of Radiation in the laser or microwave amplification by the Stimulated Emission of Radiation in the maser are two systems which are nothing more than mere resultants of very basic components.

Chapter 2
Electromagnetic Wave Theory

To gain a better understanding of the laser and the maser, a working familiarity with the science of light and propagation is necessary. To understand light, you must first have an understanding of *electromagnetic waves*. Light waves are essentially just an extension of electromagnetic waves.

FREQUENCY

Waves, frequency and the electromagnetic spectrum can be explained graphically as seen in Fig. 2-1. In Fig. 2-1A there is a drawing of a wave. In Fig. 2-1B there are a number of complete waves with the distances between the crests or peaks of each wave marked.

It took just one full second to draw this wavy line. Since two waves were created or propagated in one second, the propagated wave has a frequency of *2*.

Frequency tells us how frequently something happens over any given period of time. The distance or lapse measured in inches or centimeters between peaks of the wave forms are referred to as *wave lengths*. They can be measured directly with a ruler placed over the peaks.

The propagation of the wavy line consists of three peaks forming two complete waves that alternate above and below the centerline. Counting only the peaks of these waves, two

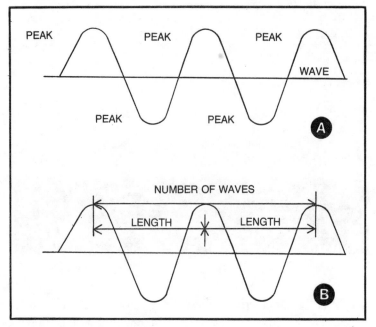

Fig. 2-1. An electromagnetic wave.

complete up and down waves were drawn in a span of one second. The artist's hand and pen moved at a frequency of two peaks, or cycles, per second. The length between waves is 2 inches, therefore, the wave is a propagation having a frequency of 2 and a wave length of 2 inches.

The notation *cycles per second* has been changed to the term *Hertz per second,* or just plain *Hertz* (Hz), as a lasting tribute to the famous scientist Heindrich Hertz. Hertz is known as the Father of the Electromagnetic Wave Theory.

Moving the pen twice as fast as before causes four complete waves or alterations to be drawn in he same one second time span (Fig. 2-2). This has propagated a frequency of 4 Hz and a wave length of only 1 inch since twice as many cycles (Hz) were drawn into the same space as had been done in Fig. 2-1.

We have now covered frequency and have agreed to refer to the number of occurrences as X amount of Hz. Additionally, wave length has been defined as the distance in inches or centimeters between such occurrences.

Almost any series of events may be drawn or graphically represented. First you have to establish just exactly what is happening and which events you wish to record.

Electromagnetic waves travel at the speed of light. The speed of light is 186,000 miles per second, or 300,000,000 meters per second in a vacuum. These E.M. Waves are composed of fluctuations similar to the waves in Figs. 2-1 and 2-2.

Each oscillation of a wave shows the energy that the wave carries. The magnitude of the rising and falling, or value of energy, is very predictable. It tells us how many times this rise and fall of energy occurs in any given time. It also gives us a direct mathematical read out of the wave length in so many units.

This self identifying phenomenon only requires the speed, number of cycles Hz and the amplitude per cycle before a graphic picture of it can be drawn.

All E.M. Waves travel at the same speed—the speed of light (186,000 miles per second or 3×10^8 meters per second).

An example of and E.M. Wave having 6 complete cycles per second is found in Fig. 2-3. In one second this wave has spanded 186,000 miles. To determine its wave length, divide the number of cycles into the distance it spans in one second. Thus

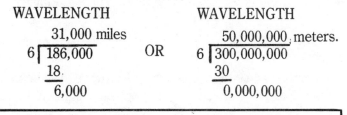

Fig. 2-2. A wavelength of 1 inch and a frequency of 4 Hz.

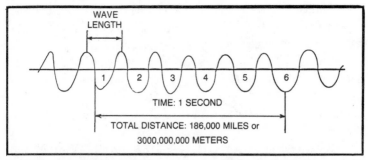

Fig. 2-3. This electromagnetic wavelength travelled 186,000 miles in one second.

However, E.M. Wavelengths are not nearly as long as our example illustrates.

In actuality, E.M. Waves start at the lower end of the spectrum or scale at about 10,000 Hz. This gives them a wavelength of 18.6 miles continuing all the way up to the extreme end of the spectrum. At this extreme right there exists micro-micro waves. These are also known as *cosmic waves* and they occur up to one-million Hz. They have a proportionate wavelength of approximately .00000000000000000000000000000001 of a centimeter in length (10^{-31} cm).

We may also make our own computations of wavelength and frequency by employing this rather simple formula:

$$C = F$$

Where C equals the speed of light (3×10^8 meters per second.)

Where F equals the frequency in Hz (cycles per second)

λ the Greek letter lamda, denoting wavelength in meters

The following example shows the length a wave of a radio station would be in meters if transmitting on a frequency of 500 KHz (500,000 cycles per second).

$$C = 3 \times 10^8 \text{ meters per second}$$
$$F = 500,000 \text{ Hz}$$

Solve for λ

$$C = F\lambda \quad \lambda = \frac{c}{F}$$

$$\lambda = \frac{3 \times 10^8}{5 \times 10^5} = 6 \times 10^2 \text{ or } 600 \text{ meters.}$$

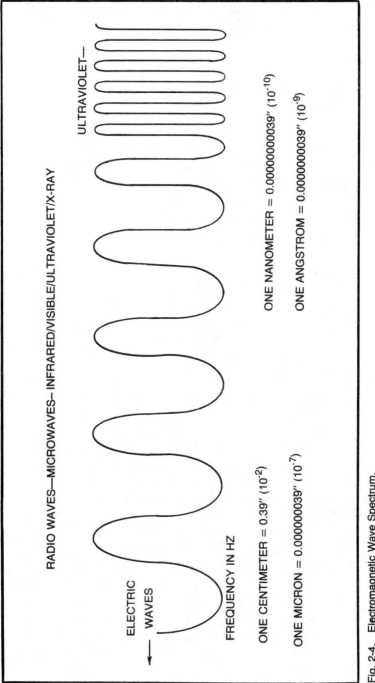

Fig. 2-4.   Electromagnetic Wave Spectrum.

27

In the preceding discussion on E.M. Waves, we have for the most part merely examined the footprints of this particular form of energy. A more working investigation into the immediate effects and potential of E.M. Waves is now necessary.

ELECTROMAGNETIC WAVE SPECTRUM

The *Electromagnetic Wave Spectrum* (Figs. 2-4 and 2-5) gives a graphic illustration of the apparent continuity of the wave-frequency relationship. It is propagated from the longer length radio waves at the far left to the higher frequency, shorter wavelength regions at the far right. The interconvertible units of length have been included as a reference because we'll be using them later in our calculations. We will be dealing with unusually minute lengths such as the *angstrom*, the *micron* and the *nanometer*.

Paving the way for the further study of E.M. Waves requires a summation of that which has thus far been covered.

- All electromagnetic waves travel through space at the speed of light which is 186,000 miles per second, or its metric equivalent of 300,000,000 meters per second.
- E.M. Waves differ in their frequency which affects their wavelength. Each is a mathematic function of the other.
- E.M. Waves differ in their effective strength and amplitude. This can be represented graphically by the waves, vertical distance above and below the "O" reference line. This facet of the waves' personality is a direct indication of its energy content.
- The frequency of the wave, as it vibrates, or oscillates its way through space is, under all conditions, of an identical frequency as the oscillating electric charge which propagated it in the first place. Thus, if an oscillating crystal were operating at 450 KHz, the wave it would propagate into space would also be oscillating at 450 KHz. It would require a tuning circuit capable of resonance at exactly 450 KHz for it to be received and discriminately detected by electronic means.
- All E.M. Waves are similar in properties. This means that one wave is not absolutely peculiar to the

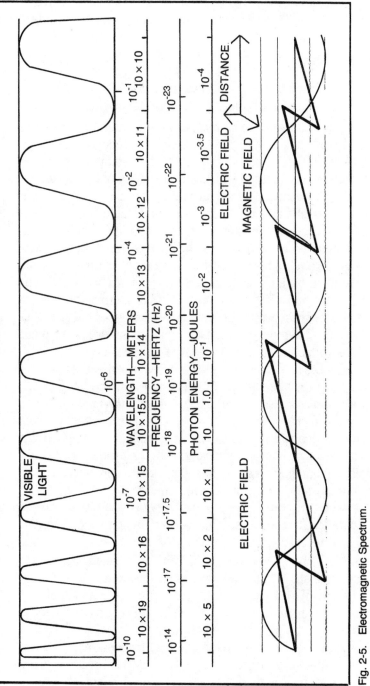

Fig. 2-5. Electromagnetic Spectrum.

29

device which generated it. Thus, an E.M. Wave of say 200 KHz propagated by the piezoelectric effect of a quartz crystal would be characteristically identical to an E.M. Wave generated by an electronic *tank circuit*. It matters not by what manner or *mode* the wave is generated by. An E.M. Wave is an E.M. Wave, regardless of its parental origin.

Any of the aforementioned devices or methods are capable of producing E.M. Waves of some frequency. The energy level may differ, but characteristically its personality is created by any method which has the ability of propagation.

The transmissions and final reception of E.M. Waves are in all instances, reversible and interconvertible between sender and receiver and between receiver and sender. This means that an E.M. Wave can be created electrically, transmitted through space, received in a resonant tuning circuit, transferred back through space to the original source of generation and still maintain its exact physical personality in every detail. Of course, it may lose some of its original energy level which may have become exhausted in transit or lost through the reproduction of it in the transferring devices.

All E.M. Waves are in themselves a manifestation of energy. They were at one time created or propagated either by a *piezoelectrically* vibrating crystal, or a *thermionically* hot filament, or the discharge of an electronic capacitor. It could also be propagated by an electron dropping from a higher to a lower level of photoelectronic energy or by the alternating field surrounding an electromagnetic field. This would represent a percentage of that applied energy.

The applied energy is represented by the waves' *amplitude*. It is a measurable vertical height or distance in either direction, perpendicular to the *time line*.

Until now, we have considered but the singular occurrence, or instance of one wave by its lonesome.

E.M. Waves can, and quite often do, exist in multiples at any one given instant. They can all be of the same frequency, or they can be different frequencies.

If they exist as waves of different frequencies, they can be segregated by their respective Hz. They will be individually tuned and dealt with independently of each other.

When more than one E.M. Wave occurs simultaneously and has the same frequency (Hz), it is to be dealt with quantitatively. It must be determined if these waves are *in phase* with each other or out of phase. If they are, it must be determined by how much and expressed in degrees of cycle. One full cycle equals 360 degrees, a half cycle equals 180 degrees, a quarter cycle equals 90 degrees, etc.

The graphically expressed simultaneous occurrence is called *not phase* and is a multiple cycle of waves. This shows the difference in the *lag-lead* relationship between the two or more occurrences in respect to their starting and stopping points, or points of interception along the time line of a wave study graph.

To satisfy immediate and practical interests, attention should be turned to waves of the same frequencey (Hz). Waves of the same frequency include all of the waves of *red light* which has an Hz of 4.3×10^{14}. Waves of red light exist only at this frequency, but more often than not, do exist in groups of waves which are either totally or partially out of phase with each other.

AMPLIFICATION

When more than one wave occurs simultaneously and in phase with its group exactly, it has an *amplifying effect*. All of the values of the waves add up to the effective value of a much greater wave. Thus, many smaller values could be amplified, providing the sum of all of the smaller values were in phase with themselves. This phenomena is exploited to a great extent in the amplification of Light, and microwaves, which we shall witness in later chapters.

Examine the two waves that exist simultaneously and in the same frequency (Hz) in Fig. 2-6. The only difference is

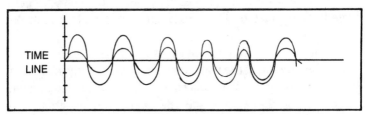

Fig. 2-6. In phase electromagnetic waves that are additively superposed.

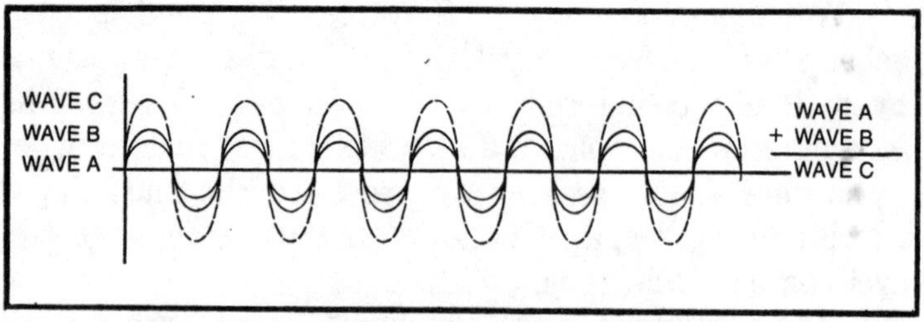

Fig. 2-7. Wave C is the resultant wave of waves A and B.

that one wave has a greater amplitude than the other. But they are in phase, exactly. We could also say that they are super-posed. Being in phase means they are additively superposed.

When one or more waves exist, they are said to be superposed. It is irrelevant whether or not they have the same frequency.

In the preceding examples and commentaries of Figs. 2-6, 2-7 and 2-8 superposed waves were considered. They were in the additive state, which would produce a resultant wave of proportionately amplified value. Superposed waves of a subtractive nature were also considered. One or more waves had opposite and exactly opposite amplitudes. Their components tend to subtract or cancel the resulting effect of each other in the productive or resultant sense. Bear in mind that as long as the waves are either *exactly in phase* or *exactly out of phase*, their resultant or effective amplitude is the algebraic collection of their respective plus or minus values. If the components are all plus, or of a positive amplitude, they will proportionately amplify the resultant wave. If they are a combination of plus or negative values, then they will produce an interference effect. This cancellative resultant will be less than the otherwise mathematical sum of the individual compo-nent values.

Only those waves of either in phase or out of phase values have been evaluated. Their extremities were either in or out of their phase.

The situation can also present itself, as it so frequently does where some of the waves are not quite exactly in or out of phase, whereby the wave will have a phase difference or angle of somewhere between 0 and 180 degrees. This situation is handled just as adeptly by the use of phasors in drawing a resultant parallelogram. The resultant is found as the diagonal is formed by the phasors, or component length sides of the constructed *PHASOR BOX*. Plans for the construction of phasor parallelograms have been included in chapter 9 of this *Laser Experimenter's Handbook*. However, it quite practically serves our immediate and instructional purposes to limit our considerations right now to the extent that waves exist either in phase or out of phase. Their individual or component values are merely added together. One wave with a value of one unit and another of two units equals three when added algebraically. All the numbers should be positive since they will be measured from the top of the time line.

The resultant sum of our waves is a product all of its own, so to speak. It is calculated as though it were one resultant wave. A resultant wave exists as shown by its effects as a resultant quantity of *amplified energy* (Fig. 2-7).

If there are two waves in phase, exactly, they exert an additive effect on the resultant wave of their creation. By this additive action we say that the waves or the amplitudes for which they represent have an amplifying effect on the resultant wave which previously and independently existed as a superposed component of.

In any case, the resultant wave of two or more superposed component waves will yield by the algebraic collection of their individual values (amplitudes), a resultant affect either greater than, or less than the sum of its components. This will

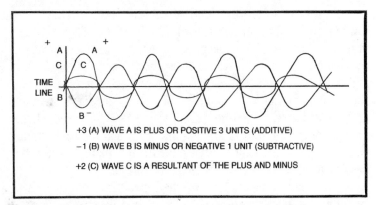

+3 (A) WAVE A IS PLUS OR POSITIVE 3 UNITS (ADDITIVE)

−1 (B) WAVE B IS MINUS OR NEGATIVE 1 UNIT (SUBTRACTIVE)

+2 (C) WAVE C IS A RESULTANT OF THE PLUS AND MINUS

Fig. 2-8. Electromagnetic waves in exact out of phase subtractive superposition.

depend on whether its components are additive (reinforcing) or subtractive (cancellative).

The effects of superposed component waves are of an additive, or reinforcing nature. They have computed their resultant effects by the newly created resultant or amplified wave which their reinforcement has produced. This represents a sum of the individual amplitudes, or a resultant of their components.

We could as well have the opposite effect incurred by having two or more waves occurring simultaneously. They will be of the same frequency, but of an *out of phase* relationship. Waves which are *exactly out of phase* are out by 180 degrees. Thus in fact, while one wave is seen in its upswing, the other wave is in its opposite down-swing. Both waves behave absolutely in contradiction of each other.

The term exactly out of phase can also be defined as an effort by two or more waves to attempt to exist simultaneously in the same instant of time. They will do so while having amplitudes that are exactly opposite and subtractive to each, or all of each other's values. In attempting to co-exist with their subtractive components, they are ultimately *cancelling* the respective effect of the other's values. Because of this joint cancelling effect, they will, by their algebraic collection, produce or yield a resultant of less than their equivalent face value.

EVALUATION OF ELECTROMAGNETIC WAVE THEORIES

In this chapter on electromagnetic wave theory, the fundamental characteristics of E.M. Waves have been studied. They were explained in respect to their graphic representation, their physical construction and nomenclature and most importantly, their proven nature to exist or manifest themselves in multiple quantities. By so doing, they exert an additive and amplifying effect on the resultantly created wave. Or in an opposing subtractive manner by which the component waves behave in a cancellative nature, the resultant effect is caused by the concerted, but interfering values.

The reader has seen mentioned, repetitiously, the term *amplify,* and *amplification of,* which I emphasized at appropriate times where the definition of amplification was being constructed. It is by this process of amplification that our lasers and masers are able to perform their extraordinary feats, and it is by the action of superposition of in phase energy waves that amplification occurs. We have then reached another level in our progress into the study of the laser.

We have now a conversant knowledge of amplification, which is a characteristic property of electromagnetic waves. The journey to the heart of the system will continue by exploring the component branches of physics which produce the resultant laser and maser. After examining enough of the components, we will consider the constructed product. But before then, there are several more concepts or components to deal with. The next one is the *light wave theory*, where much of our recently acquired experience in E.M. Wave theory will be utilized to good purpose.

Chapter 3
Light Wave Theory

In chapter 2, we were introduced to the *electromagnetic wave spectrum*. This spectrum, or *scale* was divided into several *regions* or families. One such region is the family of light, including the submembers of infrared, visible and ultraviolet. We will deal exclusively with the family of light in this chapter.

Ordinarily, when we think of light, we think of that which we can see or sense. Put more scientifically, those E.M. Waves propagated between the frequencies of from $10^{\times 14}$ to $10^{\times 15}$ and having respective wavelengths of from 0.00007 cm. to 0.00004 cm. give us the impression of *visible* colors when received by the eyes' photosensitive retinae. For example, the color red has a wavelength of 0.000066 cm. and the color violet has a wavelength of 0.000042 cm. Before proceeding further, study Table 3-1 for purposes of color orientation.

It is a fact then, that every color is created by its own particular identifying wavelength and frequency. White light, or that light given off by the sun or an ordinary lightbulb of the incandescent variety, is a representation of *all the possible colors at once*, or a combination of all of the individual frequencies. The total effect is that no particular color or frequency stands by itself, or predominates, thus resulting in seeing or sensing all of the colors—the entire mixture of wavelengths simultaneously. Again, when a source radiates or emits all

Table 3-1. Wavelengths of Visible Light.

Wavelengths of Visible Light.						
INFRARED						UNTRAVIOLET
	NANOMETERS					
1200	700	600	550	500	400	200
	CENTIMETERS					
0.00012	0.00007	0.00006	0.00005	0.00004	0.00004	0.00002

INFRARED ** RED ** YELLOW ** GREEN ** BLUE ** VIOLET ** ULTRAVIOLET

COLOR	WAVELENGTHS
RED	0.000066 cm. 660 nm.
ORANGE	0.000061 cm. 610 nm.
YELLOW	0.000058 cm. 580 nm.
GREEN	0.000054 cm. 540 nm.
BLUE	0.000046 cm. 460 nm.
VIOLET	0.000042 cm. 420 nm.

possible wavelength colors simultaneously, thereby not permitting the eye to single out or discriminate any one particular wavelength or color, we have white light, or *polychromatic light*. The chromatic situation is as follows: polychromatic light means many colors and monochromatic light means only one color.

The family of light waves, including infrared waves, visible waves and ultraviolet waves, possess, in addition to their qualities as general E.M. Waves, the following abilities or characteristics:

- They are *Rectilinear* in their propagated direction, which tells us they are existing as *rays of energy*, behaving as if they were straight lines.
- They are able to undergo *refraction*, or be bent, and have their direction diverted as they *pass through* a medium which would offer optical resistance, or a change in optical density. They do so in a very prescribed or predictable manner, resulting from a difference in speed readjustment.
- They are capable of *interference*, the function of superimposing their presence on other nearby waves, and thus affecting itself and its neighboring wave.
- They are *reflective*, or able to be reflected from a boundary, or surface of a medium. They do so under the strictest of mathematical precision.

■ They are convertible in that their energy may be converted or transferred from its original manifestation or existence of light to that of another form, such as heat, either electronical or chemical. The existence can also be changed by the disturbance of molecules to an extent which could either impart their energy or convert it in whole to any other manifestation of observable significance.

Each of these functions and characteristics will be treated individually, but first there is yet another facet of light to examine.

Coherent or *Incoherent* light is the light which we receive from the sun or from an incandescent lightbulb. The light is radiated or sent out in all directions and consists of all manners of polychromatic waves. This all occurs simultaneously and interferes with itself. Because of this diversion, or spreading out, there is a loss of brightness and intensity as it progresses from its source to its final point of destination.

If we were to eliminate all of the complementary colors from this gross and random bunch of multiple frequency rays or emissions, and leave just one primary color, we would have a monochromatic, but still *incoherent* collection of light. Totally polychromatic light is basically composed of the three primary colors of red, green and blue. Thus, we could have either a red, blue or green incoherent light beam.

If we were to further reduce this monochromatic beam of light by reflecting it with a parabolic reflector into a parallel stream of rays, we would then have a monochromatic beam of light. It could then penetrate its way into a medium further than when it was just diverged into space. An example of this type of light is that which is produced by a search light.

Taking this monochromatic light and arranging all of the individual rays of its composite beam in such a way as to cause all of the constituent and fundamental waves to vibrate, or oscillate in step with itself, will produce *coherent light*. The component waves are then constructively reinforcing each neighboring wave.

Coherent light is therefore, as far as the laser is concerned, monochromatic in phase emissions, or rays of radiant energy that are emitted in a beam like manner. Because it is coherent, it is amplified in relation to the total of *in phase* emissions. Since by virtue of this *in phase* personality, the total affect is one which would impart constructive character to the finished product. In this case, the constructive character would be the beam of emitted coherent light.

Now that the character or identity of light has been examined, let us see how it will behave under the affects of refraction and reflection. This is nothing more than a situational obstacle course for our light rays.

REFRACTION

The actual scientific definition of refraction requires a lengthy description. Its actual and specific interpretation far exceeds the mere practical value it will have for us as experimenters. Due to my commitment of practical brevity, I will provide the reader with only a working familiarity of the concept of refraction and provide a more detailed elaboration on the subject later at the advanced level. *Refraction* can be defined as the property the light waves exhibit when *passing through* a medium or substance, such as a glass lens. Due to the difference of the lens in optical density, the wave will bend, or refract from its original or incidental line of travel or direction to a new and angularly different line of exit. The exact angle of its *exit* compared to that angle of its *entrance* into the medium of different densities is a function of the medium's refractive ability. It is expressed as a percentage of the reduction in speed it can produce on the light wave as opposed to its normal speed through a vacuum, which is unity (1.0). The exiting angle or degree of refraction is also directly dependent upon the *incident*, or original angle as drawn to a *line normal* on the surface of the lens.

As light waves meet the surface of the lens and enter it (Fig. 3-1), the line of direction is *bent* upon its arrival into the refracting medium. As it progresses through the lens, it breaks through the exiting surface and its line of direction is again bent to yet another angle. This refracting ability of any lens is caused by the medium's retarding capacity. This slows down the speed of light while it passes through the medium. It

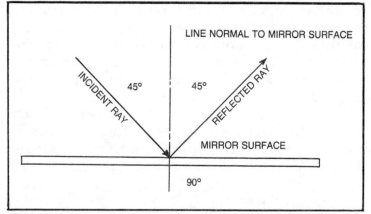

Fig. 3-1. Refraction angles of incident rays entering and exiting from a converging lens.

is the prompt and deliberate change in velocity that causes the break in the light rays' direction of travel. This turning or bending is not an external function, as in the case of *reflection*, where the phenomenon is more analogous to a ball bouncing off a surface. *Refraction* is a different concept in that it is achieved internally to the medium, and not externally. An example is that which takes place with a mirror, where the rays never actually pass through the medium, but interact upon the external surface.

REFLECTION

Reflection can be compared to a wave striking or being *incident* to a surface in a very predictable manner. The wave bounces or reflects off of the *incidental* or reflective surface. More specifically, the *incident wave*, or particle, will be reflected by a *reflecting surface* at the same angle of its arrival. It is measured in respect to a line drawn normal to its surface (Fig. 3-2).

For the purpose of simplicity, Fig. 3-2 describes a nominal angle of 45 degrees incident and an equal reflection. This angle, in respect to the *line normal* of its surface, could be any angle at all. Thus, if the incidental ray were to arrive at the reflecting surface at an angle of say 36.8 degrees *to* its *line normal*, then it would be also reflected at 36.8 degrees *from* its *line normal*.

The term *line normal* should be committed to memory by the reader, because it will appear regularly in chapter 4 and will be employed in our practical applications when we begin the actual construction and experimentation with lenses and mirrors. It should also be mentioned that our mirrors will not be just flat surfaces, but rather of the spherical and parabolic varieties.

The functions and behavior of our light rays will, regardless of the geometry of the mirrors' reflective surface, obey all the laws and mathematical rules laid down for them in basic flat or straight situations.

We will now treat the *cause* of the actions of refraction and reflection. One such cause is the fashion in which our light rays travel either through a medium or a vacuum, that is, in a straight or rectilinear manner.

Rectilinear Propagation is one of the foremost problems encountered by our founding scientists in their exhaustive quests for the explanations of light wave phenomena. It is the ability and manner in which light can manifest itself by existing as both *traverse waves* and as individual particles of energy.

Having arrived at the conclusion that light rays do exist as particles, or bullet-like manifestations, we can now theorize

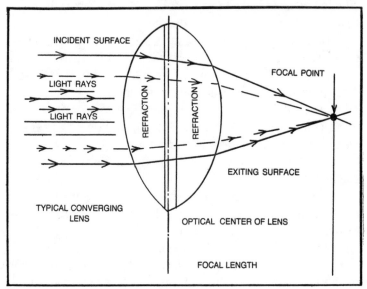

Fig. 3-2. An angle of incidence and reflection.

further and postulate certain easily understood functions the rays can perform.

Light rays are rectilinear in their travels or propagation. Consequently they form straight lines and we may apply direct geometry in calculating their reactions to, and interactions with, any medium or surface which would interrupt their intended or incidental path of travel. Examples include the surface of a mirror, or the internal and external boundary surfaces of a refracting lens or some other transparent medium.

We may, because of the rectilineal straightness of rays and the inherently geometric advantages of description, design mechanisms for their use and do so with mathematical precision. We can predict the reactive angles from a straight line course because light rays are rectilineal, almost as if they were exact lines which could be drawn with a pen or pencil. We can utilize this phenomena to great advantage when designing mirrors and lenses and their functions.

INTERFERENCE

We will consider lights' character of *interference*, which for our purposes is to be evaluated only as the resultant affect upon the waves' intensity in respect to the *energy level*. It is affected while in transit from its point of emission or propagation as well as at its point of destination.

Due to the existence of multiple simultaneous waves, interference is the action by which one or more waves will tend to cancel the value of another wave's effective amplitude. The co-existing waves will exert an additive influence on each other and produce an effective wave of greater amplitude than of its components. Or, the co-existing waves will interfere with each other in a subtractive manner which would then have a destructive effect on the final product of the waves' sum of components.

Interference occurs in all instances where we have the co-existences of multiple waves. The only difference is in the manner they exert their influences. The effect of *constructive interference* is in the amplification of the emission as a whole. The effect of *destructive interference* results in the diminution of the emission as a whole. Interference exists constantly, whenever two or more waves get together.

Interference also has an influential function on the chromatic effect of E.M. Waves. In addition to the fact that light rays are being dealt with in the particle context, they are first and always, E.M. Waves. They are affected by any phenomena which affects their ancestorial genetics as extended families of the E.M. Wave Spectrum. Light waves are directed toward their energy transmitting character.

The aspects of reflection, refraction, rectilinear propagation and interference have all been discussed.

We can now describe the laser beam. It is an emission of *collimated* (parallel, pencil-like, non-diverging rays) and *coherent* (all waves being in phase with themselves) appearances of light energy. These appearances are in either the infrared, visible or ultraviolet regions of the E.M. Wave Spectrum.

This appearance is visual if it occurs between 10^{14} and 10^{15} Hz. It is infrared if it is non-visible and occurs below 10^{14} Hz. Ultraviolet (above 10^{15} Hz) will monochromatically identify itself as to its propagated frequency. We know that the chroma is directly indicative of its frequency and corresponding wavelength. In the event that the emission is non-visible, the frequency may be equally identified by infrared and ultraviolet measuring instruments.

A light amplification by the stimulated emission of radiation is, in effect, spelling out the very subject of this book.

In chapter 5, we will conveniently assume that it is just such a laser beam that we are putting through the paces. Observe the proficiency with which it accomplishes its tasks and behaves at our instant disposal.

Chapter 4
Mirrors and Lenses

As seen in the preceding chapters, light rays, or E.M. Waves can be made to perform a number of prescribed tricks. Certain characteristic functions may be utilized to affect certain phenomena by the use of mirrors or lenses.

One such characteristic of light, as was described in chapter 3, is the inherent properties of reflection and refraction. Both of these may be proven and exploited by the use of mirrors and lenses, respectively. Following this order we shall first describe the function of mirrors.

MIRRORS

There are three basic types of mirrors: a *flat mirror*, a *spherical mirror* and a *parabolic mirror*. The flat mirror (Fig. 4-1) is probably the most widely known type in common use. This is the very same type we look into every morning. It consists of a flat piece of glass which is silvered on the back side. It is capable of reflecting light on the opposite side, its reflective side.

It will be noted in Fig. 4-1 that the *line normal* to the flat mirror lies at exactly right angles (90 degrees) from the plane of the mirror. As stated in chapter 3 the incidental and reflected rays of light are measured from this line normal.

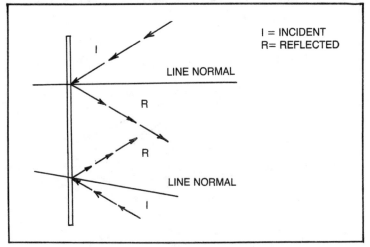

Fig. 4-1. A flat mirror. The line normal lies at a right angle to the plane.

A spherical mirror (Fig. 4-2) is physically a half of a solid sphere and silvered on the outside. It is reflective on the inside, which is its concave side.

A parabolic mirror (Fig. 4-3) is almost exclusively used in search light reflectors and microwave transmission antennae. Although no specific use of this type of mirror is employed in the construction of the laser or maser, it was mentioned to

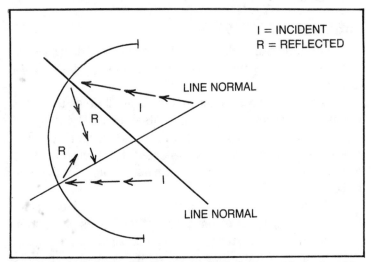

Fig. 4-2. A spherical mirror. The line normal equals the geometrical radius of the mirror.

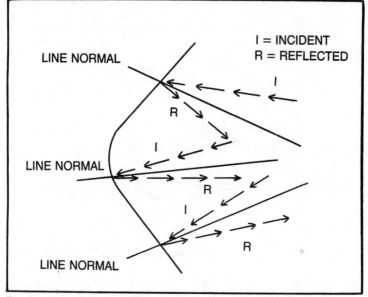

Fig. 4-3. A parabolic mirror. The line normal equals the line of parabola.

provide the reader with at least the knowledge of its existence.

Since spherical mirrors are the most widely used reflector found in our system's construction, this variation will be disussed at length.

THE SPHERICAL MIRROR

The important reflective and constructional data and facts related to spherical mirrors can be found in Figs. 4-4, 4-5, 4-6 and 4-7.

The line normal for this mirror is at all times the actual radius used to develop the sphere itself (Figs. 4-4 and 4-5). Consequently, all arriving incident rays falling upon its reflective surface will be accordingly reflected at an angle equal to its incident arrival. This is measured from the line normal—the radius of curvature.

Because of its concave curvature, the reflective rays will have the ability to meet, or focalize themselves at a certain point or distance from the mirror. This is called the focal point.

If certain parameters of curvature geometry, the distance of light source from the mirror, and an area of light source were to meet, all of the light rays would be returned or reflected back to the point of creation, called the *image*.

This spherical mirror (Fig. 4-5) is shown as a cut-away. It also illustrates the *radius of curvature*, which is also the line normal and the *focal length*, which is the distance that the rays will meet and the *principal*, which is a point from which the radius of curvature is centered.

Observe that the random angles of incident rays, although obeying the prescribed laws of reflection, are not meeting or focalizing at any one particular point (Fig. 4-4). This is because of the lack of singleness of projection. All of the rays could be focalized to a definite point of destination or crossing were such light sources maintained for the issuance of the light rays (Fig. 4-5).

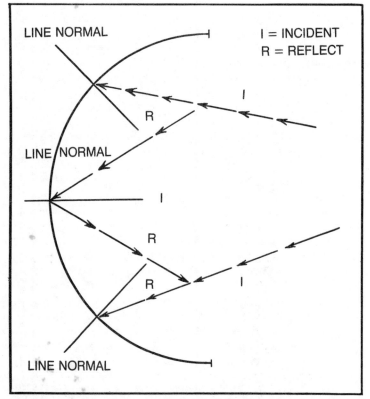

Fig. 4-4. A spherical mirror illustrating the geometry of incident light rays and reflected light rays to the line normal of the mirror. The line of reflection is always equal to the line of incidence.

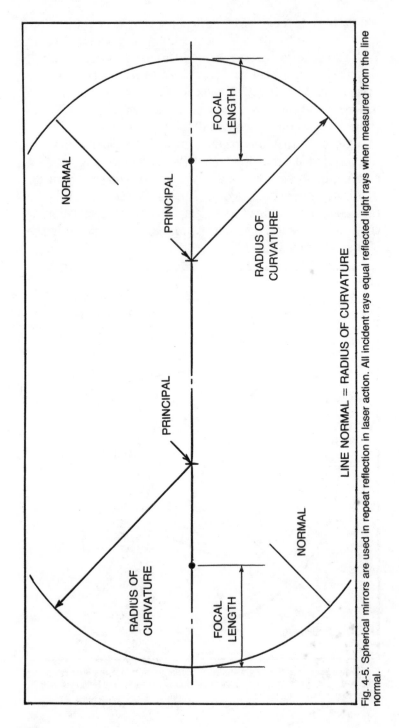

Fig. 4-5. Spherical mirrors are used in repeat reflection in laser action. All incident rays equal reflected light rays when measured from the line normal.

LINE NORMAL ≡ RADIUS OF CURVATURE

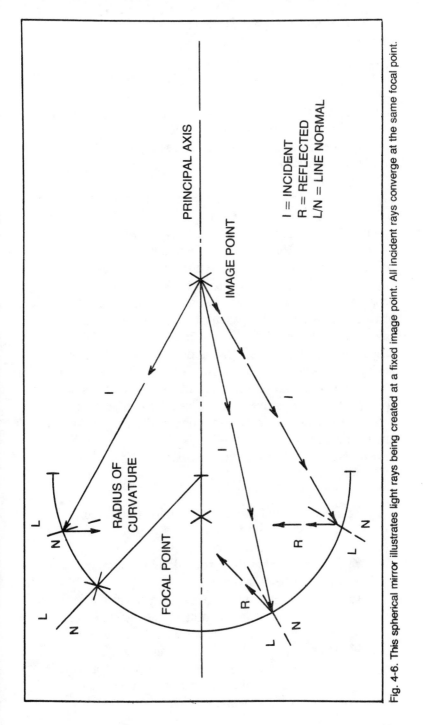

Fig. 4-6. This spherical mirror illustrates light rays being created at a fixed image point. All incident rays converge at the same focal point.

49

In Fig. 4-6, the light rays have been created and aimed toward the spherically reflective surface. The reflected rays tend to meet at one point. Were we to make use of two mirrors (Figs. 4-7 and 4-8), we could adjust the point of image and curvature of reflection to reflect the rays of one mirror which would serve as the image for the second mirror. The second mirror would then return the rays to the first. This receive and return process would continue as long as a light source were to provide the light necessary for repetitive reflection (Fig. 4-9).

An internal action is caused by the repetitive reflected rays that occur when a *lasing* or *masing* action is in progress (Fig. 4-10). The rays are reflected between the two mirrors, A and B. These reflectors form the cavity end walls of the laser or maser. Mirror B is deliberately made a little less reflective than the opposite end mirror (A). This provides a window for the exit of the produced laser beam, or maser propagation. In the early gas lasers, the mirrors used were placed externally to the gas tube at each end. Due to the advances in glass blowing, the two mirrors can now be sealed inside the gas filled tube. This is done after being properly designed and spaced at appropriate distances from each other. This trend allows for the convenient assembly of gas lasers (Fig. 4-11) and experimentation. This is a tremendous benefit since previously, the least bump or jar would result in just enough misalignment to render the laser unfunctional. Needless to say, this caused many associated headaches for the experimenter.

In summary and reiteration:

- Regardless of the particular mirror selected, a line normal has been inherently designed into it by virtue of its respective geometry.
- A flat mirror has a line normal which is at right angles to its plane of surface.
- A spherical mirror has a line normal which coincides with its radius of curvature. The radius has been centered from this point, called the principal.
- All angles of incidence and all angles of reflection are equal. They are measured in degrees from and in relation to the line normal of whatever particular mirror is under consideration.

■ The reflective characteristics or focal points of spherical mirrors may be altered to suit the specific requirements of the system being designed. This can be done by changing either the radius of curvature of the mirror itself or the distance between the mirrors.

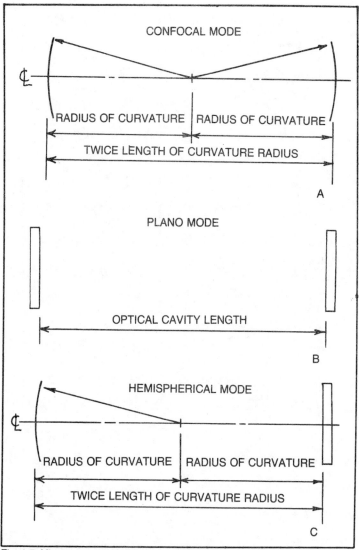

Fig. 4-7. Mirrors are used to construct optical cavities. A. Two spherical mirrors. B. Two plano (flat) mirrors. C. One spherical and one plano mirror.

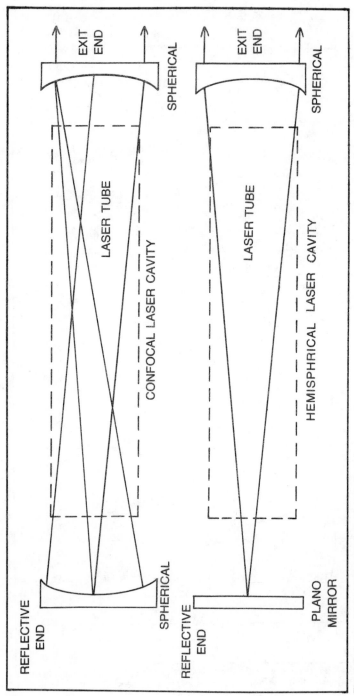

Fig. 4-8. Laser cavity showing mirror arrangements.

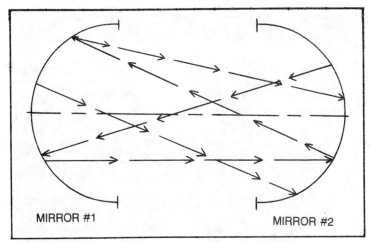

Fig 4-9. The repetitive reflection of two mirrors.

LENSES

As mirrors are an optical device by which the properties of reflection may be exploited, it may be conversely stated that lenses are a device whereby the property of refraction may be utilized. A lens, regardless of the optical substance of

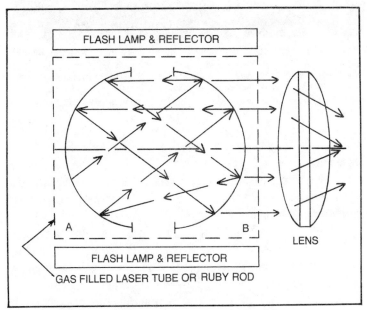

Fig. 4-10. Repetitive reflection as employed in a laser cavity.

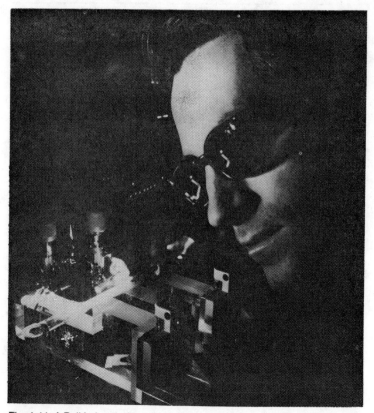

Fig. 4-11. A Bell Laboratories scientist demonstrates a miniaturized waveguide gas laser. The fine line of laser light in glass tubing (at left) is passing through a waveguide configuration with a diameter of about 20 thousandths of an inch. Such lasers can be used in communications systems that use light to carry large numbers of voice, picture and data signals. (Photo courtesy of Bell Laboratories.)

which it is composed, serves the function of refraction. It slows down the light rays entering it. This specific ability to retard the transmission of light is known as its *refraction index*. It will bend or refract the rays' direction of travel in respect to its own surface and line normal.

There are a variety of lenses available. But basically, there are two distinct types of lenses; the *concave* type (Fig. 4-12) and the *convex* type (Fig. 4-13). We will be dealing almost exclusively with the convex lens which imparts a converging effect (Fig. 4-14) on the exiting rays that pass through the refracting material.

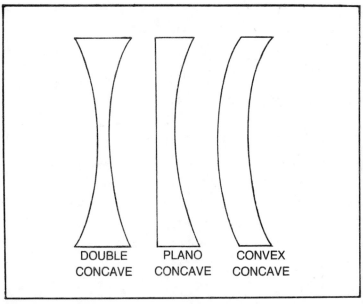

Fig. 4-12. Concave lens.

The concave type causes the exiting rays to diverge (Fig. 4-15) and has no real value in the experimentation of the systems included in this book. Therefore no further mention will be made of this type of lens.

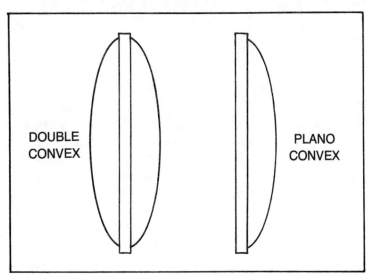

Fig. 4-13. Convex lens.

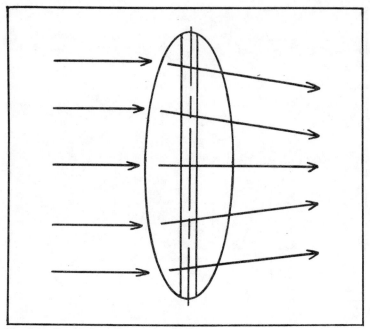

Fig. 4-14. Converging effect on a convex lens.

THE CONVEX LENS

There are several styles of convex lenses available. Each has its own special characteristics and properties for which the lens was specifically ground for. This study will be limited to the type known as the *double convex converging lens*. This lens probably has about the best features of any lens obtainable.

A *double convex lens* (Fig. 4-16) is made by grinding a suitably refractive piece of glass to a prescribed radius of curvature. This radius of curvature is similar to the radius mentioned in respect to a spherical mirror in that it has a radius extending from the point of principal at its center.

It is essential that a lens be selected that will refract the rays to a near perfect point for the purpose of focalizing the greatest portion of light rays. This would be vital if we were to build a laser for the purpose of creating a heated focal point sufficiently hot enough to melt steel. In that case, a lens would be selected that could refract the laser beam to a focal point about 4 to 10 inches from the center of the lens. This would concentrate on the collected beam and aim it at a point far enough away from the lens to keep the lens itself from melting or deforming from the intense heat.

In a CO_2 laser for example, the lens selected will be able to concentrate so much light energy at one given point that it will literally cause the air to glow with incandescence. It will melt the hardest of steels and even burn holes in diamonds. This is being done on an industrial scale. By selecting a mild laser and a different lens arrangement, the human eye may be surgically operated on without even opening the eye. This *Retinae Welding* has been in practice now several years. Which extremities will illustrate the flexibility and adaptability of the laser depends upon the power output of the beam and the lens arrangement selected for the problem given.

For our purposes then, the necessary lens questions which will need answers for the project construction include:

■ Will the lens concentrate the light as with a convex type lens, or will it diverge the light as with a concave type lens?

■ What will the lens' focal point be?

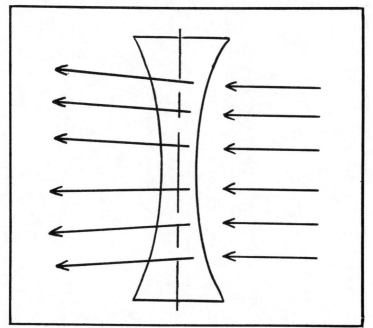

Fig. 4-15. Diverging effect on a concave lens.

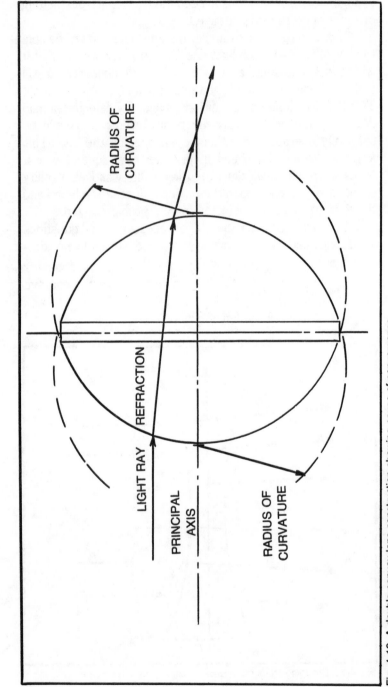

RADIUS OF
CURVATURE

RADIUS OF
CURVATURE

REFRACTION

LIGHT RAY

PRINCIPAL
AXIS

Fig. 4-16. A double convex lens construction showing cause of convergence.

■ If this lens is one of two or more lenses used in teleprojection, will it be optically compatible with the component lenses?

A good book on lenses should be obtained before any extensive projects are undertaken involving special arrangements of lenses. Most science suppliers will be more than glad to provide the experimenter with a catalog of the optical manufacturer who they themselves order from. It makes filling orders easier for them that way since every manufacturer has their very own specifications and coded data pertaining to their own lenses and optical goods.

However, the information the reader has been provided with should suffice in making an intelligent survey as to just what type of lens and dimensions are needed. These should be specified in ordering supplies in respect to the properties of lenses and mirrors. Where a lens is required in the plans in this *Laser Experimenter's Handbook*, the necessary data has been nominally specified. A list of suppliers has also been included in Appendix C.

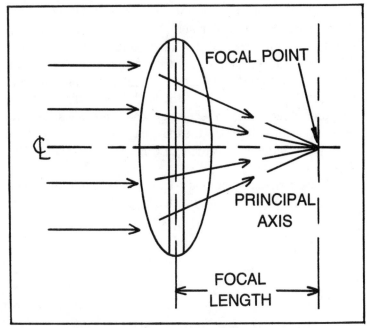

Fig. 4-17. A single lens produces this focal point for concentrating rays.

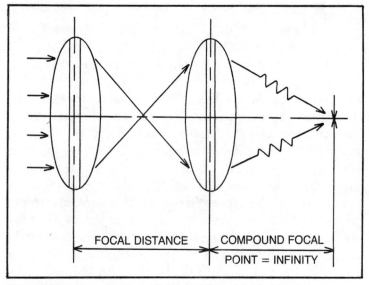

Fig. 4-18. The focal distance and compound focal point of two lenses.

Where compounding is necessary, a simple tube with suitable lens' retaining rings may be made. Procure a tube of such an inside diameter that it will accommodate the lens O.D. Plastic snap rings can be inserted in front of and behind the lens to position it. Secure the rings to the inide of the tube. The tube which houses the compound component of the lens group will have an inside diameter at the first lens end which will provide for a *slip* fit over the laser tube. This facilitating longitudinal movement affects the focal length adjustment for beam projection. It makes the exit beam projection adjustable from twice the lens' focal length (Fig. 4-17) to infinity (Fig. 4-18).

Infinity means just that—lasers have even shot their beams to the Moon and back with only negligible divergence.

Chapter 5
Laser and Maser Concepts

Light rays are capable of piercing through diamonds, vaporizing the toughest materials known to man and gentle enough to penetrate the human eye and fuse a torn retina to save him from blindness. Emanating fences of invisible microwaves forewarn us of approaching enemy missiles and may soon replace the seeing eye dog as an electronic vision system. These are only a few of the present and expected feats to be performed by our newly acquired servant—the laser, which comes from Light Amplification by Stimulated Emission of Radiation (Figs. 5-1 through 5-3).

The maser, which was first to be discovered, can be thought of as a non-optical laser. It is an electronic device very similar in anatomy to the laser, but different in that the end product is an amplified and coherently generated microwave emission. However, the end product of the laser, or conversely, the optical maser results in an emission of the light region either as visible, infrared, or ultraviolet light (Figs. 5-4 and Fig. 5-5).

Both the laser and its father the maser are electrophysical systems capable of receiving, or being acted upon by very weak signals, or inputs. This process is known as lasing. It results in a final out-put reaction to the initial stimulation or tuning. This brings about a more desirable and far more usable

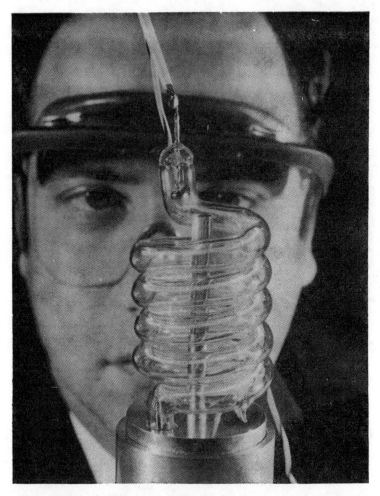

Fig. 5-1. A Hughes Research Laboratories scientist studies an early laser's main parts—a light source surrounding a rod of synthetic ruby crystal through which excited atoms generate the intense beam. (Photo courtesy of Hughes Aircraft Co.)

emission of an E.M. Wave of a coherent and amplified character which may then be capable of more sophisticated tasks. Some of these tasks can be seen in Figs. 5-6 through 5-17.

ABSORPTION AND EMISSION OF RADIATION

To understand how lasers and masers absorb and emit radiation, it is necessary to visualize the fact that atoms and molecules are not really inert units in themselves. They are to be thought of as bundles of energy. The electrons are very minute particles of electricity which are positioned around the outer envelope of the main nucleus of the atom by certain proven forces. These forces are established by the nucleus of the particular element. They can be externally affected—a principle about which lasing and masing action depends.

The atoms and molecules of solids vibrate in a very fixed position by virtue of their physical lattice. This is an architec-

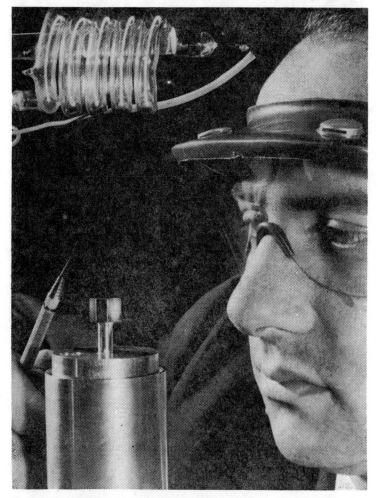

Fig. 5-2. This scientist views a cube of synthetic ruby crystal that forms the heart of a laser and generates the light. The light source (at top) is used to excite the tightly-packed atoms in the ruby which then amplify the laser's light into an intense parallel beam. (Photo courtesy of Hughes Aircraft Co.)

Fig. 5-3. This synthetic ruby crystal (top) glows with absorbed light. The light source (below) pours random waves of light into the ruby. (Photo courtesy of Hughes Aircraft Co.)

tural aspect of the crystal itself. However, the atoms and molecules of liquids and gases are more movable. They may leave their posts and wander basically without restraint within the containing boundaries of their vessel or container.

All atoms and molecules, of just about every element and compound investigated to date, have one prominent characteristic. They are affected by either light energy, magnetic forces, electrostatic or current charges, thermal influences, or in some instances mechanical reactions. This inherent personality of matter to be internally affected by external effects is utilized in both lasers and masers. It is called tuning. Actually, tuning is just another term for affecting, or influencing something. It is as if you were to tune your radio or television set to a channel. The end result is that you have affected or influenced the electronic resonance of the circuits' ability or filtering aspects. Because you did so externally to the set itself, it's referred to appropriately enough as tuning.

Atoms and molecules can exist in different energy levels, separated from one another by definite steps. Suppose that

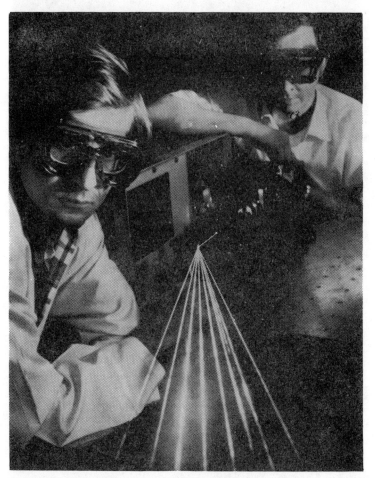

Fig 5-4. Two Bell Laboratories scientists demonstrate the range of colors—from near-ultraviolet to yellow—that can be created from a dye laser called an exciplex laser. (Photo courtesy of Los Alamos Scientific Laboratory.)

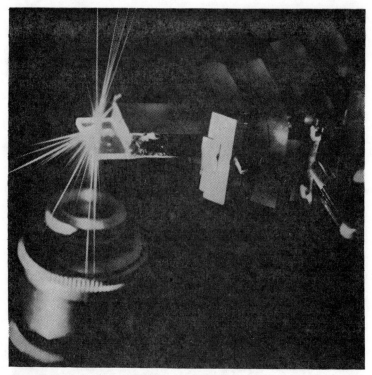

Fig. 5-5. Lasers create beams of light that range in colors.

radiation from without acts upon a substance. If a given atom of the substance is at a low energy level, it will absorb the energy. The incoming radiation therefore will be decreased by that amount. If the atom is at a high energy level when it is stimulated by radiation of the appropriate frequency, it will drop to a lower energy state. It may even do so without any specific stimulation by a wave—that is, spontaneously. Ultimately, it will emit radiation corresponding to the stimulated wave. Or, in the absence of the latter, waves will correspond in frequency to the difference between the higher and lower energy level. Where there is stimulating radiation, this will be reinforced by the additional radiation. Under ordinary conditions, most materials will absorb much of the energy falling upon them. This is because most of their atoms are in the lower energy level to begin with.

The absorption and emission of radiation is normally a disorderly and random process. For one thing, the incoming radiation is usually a mixture of frequencies. Moreover, as atoms and molecules vibrate, they interact with one another, making the energy transformations even more complex. But what if we could make the atoms behave in a more disciplined way? What if we could make them jump up and down the energy steps at the physicist's command in unison like a troop of soldiers? That is precisely what has been accomplished with the laser and maser.

Fig. 5-6. An electro-beam controlled carbon dioxide (CO_2) laser amplifier controls the same energy in the sun and starts a thermonuclear reaction. (Photo courtesy of Los Alamos Scientific Laboratory.)

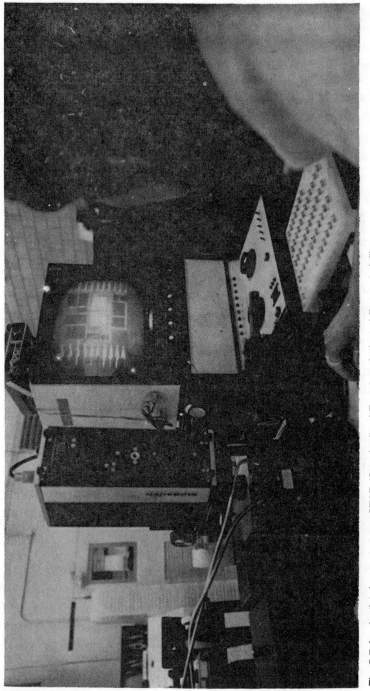

Fig. 5-7 A scientist demonstrates a STAR (Standardized Tantalum Activated Resonator) filter.

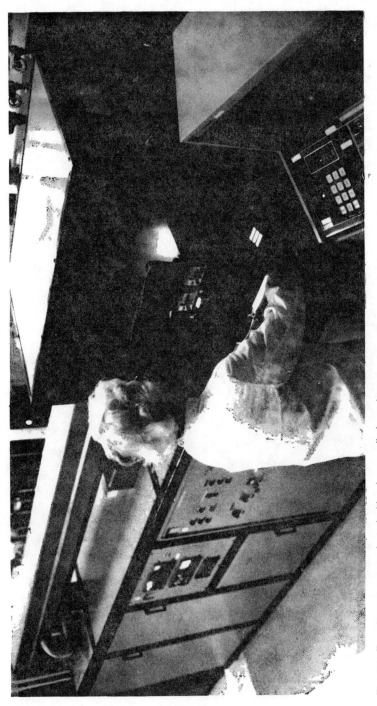

Fig. 5-8. Experiments are conducted with a laser scribnix system.

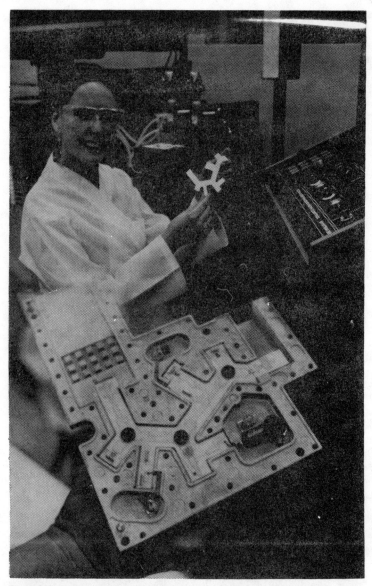

Fig. 5-9. A laser guides the light through this continuous path control system.

BASIS OF LASING AND MASING ACTION

The medium utilized in a maser can either be a gas, a solid or a liquid. The medium liquid has been selected in an engineering manner because its atomic or molecular structure has favorably arranged energy levels. It is led into, or is

70

enclosed in a cavity resonator. This is simply a container from whose walls incoming radiation can be effectively reflected. It can also be detained for a short enough period of time to affect the required properties given the energy before its release as a beam.

First, this contained lasing or masing medium will have its population of atoms or molecules raised in respect to its average level of energy. This is in contrast to that level which it normally exists in. This inversion can be brought about in several ways.

Fig. 5-10. A Bell Laboratories engineer exhibits a tiny solid-state laser perched atop a specially designed mount. Such lasers have average projected lifetimes of one million hours—about 100 years—at room temperature, according to the lab's accelerated-aging test results. (Photo courtesy of Bell Laboratories.)

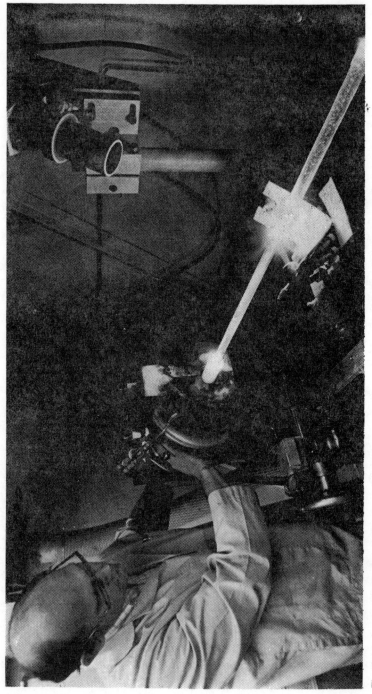

Fig. 5-11. A scientist displays the direct beam of a laser light ray. (Photo courtesy of Western Electric Co.)

Fig. 5-12. Much smaller than the grain of salt to its right this solid-state laser (small rectangle atop block) may hasten the day you talk over a beam of light. This laser may prove to be a reliable, efficient and economical source of light for a future optical communications system. (Photo courtesy of Bell Laboratories.)

In one method, the low-energy atoms are filtered out through the use of an appropriate electrical field. This was the manner employed in the first operational maser, the Amonia Maser, devised by J. P. Gordon, H. J. Zeiger and C. H. Townes.

In another method, the temperature is substantially lowered by means of liquid helium or nitrogen, thus bringing the

Fig. 5-13. Metrologic's 60-230 Photometer, in addition to being an excellent laser power meter at a very low cost, can also be used as a communications receiver. (Photo courtesy of Coherent Laser Division.)

Fig. 5-14. This collimator is a Gallilean telescope that screws into the mounting ring built into the front of a laser. (Photo courtesy of Coherent Laser Division.)

bulk (or population) of the atoms and molecules down to a low-energy state. Then these atoms are raised at the same time to the high-energy state by *pumping*. Pumping is accomplished by introducing into the system electromagnetic energy that is at a different wavelength from the stimulating wavelength that will be used later.

The next step is to introduce a stimulating wave at one wall of the cavity resonator—a microwave of the appropriate frequency. It will then excite a number of the high-energy atoms in its path. Each of them will consequently emit electromagnetic energy of the same wavelength as the stimulating wave. Thus, the energy of the stimulated atoms will be added to the energy of the original microwave. The wave will then go into resonance. It will proceed to bounce off of one cavity-resonator wall and then the other in a process that progressively excites more atoms. As the wave continues to bounce back and forth from one end of the cavity to the other, the wave is becoming progressively stronger in value. By the time it is emitted from the maser cavity, or when it breaks through the less reflective wall or mirror, it will therefore have become enormously amplified. This amplified microwave will reproduce unbelievably weak signals (E.M. Waves) with great fidelity and very little noise or interference. In the Amonia Maser only two energy states are involved. However, in most masers and lasers, the energy reduction steps are based on three or more energy levels.

Fig. 5-15. Battery-operated, solid-state photodetector and amplifier with choice of ordinary or high power outputs can act as a receiver for the beam of a modulated laser to produce an audio output when connected to an oscilloscope;..or TV picture when connected to a closed circuit monitor. (Photo courtesy of Coherent Laser Division.)

There are fundamental similarities as well as differences in the operation of masers and lasers. In the laser, atoms are raised to a high energy level state before they begin to emit. Unlike the maser however, the laser puts forth a beam of intense light. It is infrared, or ultraviolet radiation instead of microwaves that shows the different working regions of the

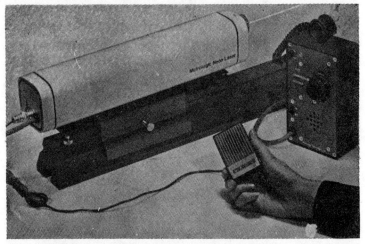

Fig. 5-16. This Metrologic Communicator Kit is designed for use with modulated lasers. The kit comes complete with a microphone which plugs into the back of the lasers. (Photo courtesy of Coherent Laser Division.)

Fig. 5-17. This scientist is using a laser for the repair of an imperfectly etched integrated circuit.

Electromagnetic Wave Spectrum. The laser does not usually function as an amplifier of a weak signal which might be impressed on it from the outside. The laser *CAN* be adapted as an amplifier of light signals coming from the outside of the proper frequency, then by pumping it to a point where it would begin to lase by itself.

Light or electrical energy is used to raise the energy level of the atoms in the laser material. The now activated high-energy atoms begin to drop back to low-energy levels, which we've learned is called decaying or de-excitation. As they do so, they begin to give off the characteristic laser radiation or beam. Since this beam is rather weak at first, it must be reflected back and forth repeatedly between its mirrors. One is at each end of its resonator cavity, as in the maser. During this resonation period the beam stimulates other high energy atoms to drop to their lower energy states, and in so doing, they will emit their contribution of energy of the same quality. This reinforces or amplifies the beam internally. When it becomes sufficiently energetic (in the merest fraction of a second), it passes out of the laser through one of the mirrors. This mirror is somewhat more transparent than the other end mirror and made less reflective. Thus we could say that this exciting mirror is the window end of the resonator. The beam of light emitted by the laser is almost absolutely monochromatic in color. This is due to its narrow range of frequency. Additionally, the beam is collimated and coherent, in reference to its non-diverging characteristic and its singleness of fre-

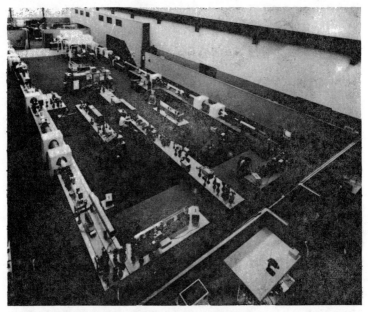

Fig. 5-18. This is an overview of the Lawrence Livermore Laboratory's two trillion watt Argus laser, one of the world's most powerful lasers. (Photo courtesy of Lawrence Livermore Laboratory.)

quency. It has no other component frequencies to interfere with itself or to reduce its own brilliance or power.

By subjecting lasing materials to various types of electrical, magnetic and even sonic energy, we can modify the laser output in various ways for the purpose of signalling, communications and general experimentation. Many different patterns can be imposed on a beam and even its color can be tuned or changed by an external tuning, such as in *tunable dye lasers*. But bear in mind that the only real difference between maser and its son the laser is its working region of the Electromagnetic Spectrum. The maser works in the microwave or non-visible regions. The laser works in the higher frequency regions of infrared, visible and ultraviolet. The maser is but an non-optical laser and conversely, we may categorize the laser as an optical maser.

LASER FUSION

There are several science laboratories that work extensively with lasers. One such place is the Lawrence Livermore Laboratory which is operated for the Energy Research and Development Administration by the University of California.

They have conducted experiments with a two trillion watt Argus laser, one of the world's most powerful and the third of a four-generation series of fusion research lasers at Livermore.

A master oscillator—the blue, L-shaped box on the center table of Fig. 5-18—generates a 10 megawatt laser pulse lasting less than a billionth of a second. It moves toward the lower right and is split in two. One pulse moves left, the other right. At the corners, they are reflected along parallel paths toward the target chamber which is barely visible behind the wall at the upper left.

During their journey from the master oscillator to the target chamber, the pulses pass through a series of amplifiers, filters and devices that protect the laser equipment from being damaged by the extremely intense light. In addition, the pulse is diagnosed several times along the way for later analysis.

The argus facility enables Livermore researchers to make studies of fusion microtargets compressed to high density. Such studies are an important step on the way to the goal of demonstrating the scientific feasibility of laser fusion.

A 1,000-joule, neodymium glass laser chain for controlled thermonuclear fusion research is also under development at the same laboratory. Increasingly powerful laser experiments will be conducted through the 1970's, culminating in 10,000-joule experiments scheduled to test the feasibility of fundamental laser-induced fusion concepts. These include the generation of micro-thermonuclear explosions in tiny heavy hydrogen pellets irradiated from all sides by extraordinarily precise powerful pulses of laser light.

Fig. 5-19. A scientist exhibits a "C" amplifier—one of the largest in the laser chain under development. (Photo courtesy of Lawrence Livermore Laboratories.)

Fig. 5-20. A 300 foot long-path, glass disk laser with a maximum energy of about 200 joules. (Photo courtesy of Lawrence Livermore Laboratories.)

Fig. 5-21. The north bank of six of the 20 Shiva laser amplifier chains looking back from the end nearest the target room. (Photo courtesy of Lawrence Livermore Laboratories.)

Fig. 5-22. Looking back from the target room along six of 20 arms belonging to the Shira Laser system. (Photo courtesy of Lawrence Livermore Laboratories.)

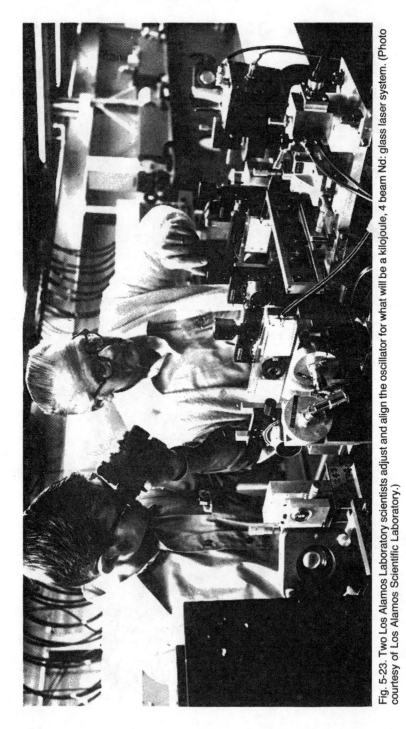

Fig. 5-23. Two Los Alamos Laboratory scientists adjust and align the oscillator for what will be a kilojoule, 4 beam Nd: glass laser system. (Photo courtesy of Los Alamos Scientific Laboratory.)

In Fig. 5-19, an assistant laser division leader opens a "C" amplifier, one of the largest in the laser chain under development. The oval disc in the center of the amplifier is made of optically superior glass doped with atoms of neodymium, a rare earth element. Near the scientist's hand are the powerful flash lamps that pump the neodymium atoms to the energy states needed for laser action.

A long-path, glass disk laser made of neodymium glass (Fig. 5-20) is also used in experiments in laser fusion at the Lawrence Livermore Laboratory. This laser is about 300 feet long and has a maximum energy of about 200 joules.

One of the world's most powerful lasers was built by the Lawrence Livermore Laboratory. One in a series of high power laser systems, it was named Shiva. Figures 5-21 and 5-22 offer a view of the north bank of six of the 20 Shiva laser amplifier chains looking back from the end nearest the target room. The Shiva laser system delivers more than 30 trillion watts of optical power in less than one billionth of a second to a tiny fusion target the size of a grain of sand.

Another laboratory that conducts laser research and experiments is the Los Alamos Scientific Laboratory. In Fig. 5-23, a laser division scientist and a physicist adjust and align the oscillator for what will be a kilojoule, 4 beam Nd: glass laser system. Preliminary experiments are being conducted with one beam while assembly of the other units is underway.

Chapter 6
Introduction to the Laser

While introducing the reader to the laser, he or she will no doubt conclude that the entire system is but a synthesis of all those individual parts, or branches of physics already studied. This is, in fact, the very point that I've tried to relate. The beginning of this chapter will really be a synoptic review to the reader, but one in which you'll be able to conceptually construct, from the ground level up, a complete functioning mechanism.

CONCEPTUAL CONSTRUCTION

In chapter 3, we discussed coherent and incoherent sources of light. As was mentioned, the atoms that make up the source of light emit this radiation in the form of photons. In an incoherent source, they are emitted in a manner called spontaneous emission. In such an emission, the photons are emitted randomly in time and direction, and have a wide range of wavelengths and frequencies (colors). Thus the light wave pattern of an incoherent source is analogous to the completely irregular wavelets produced by drops of rain water.

In contrast, the wave pattern of a coherent source, such as a laser, is like that of regularly spaced parallel water waves, being driven by a uniform wind. This regularity derives from the process of stimulated emission which occurs in a typical laser. Such an emission arises from the action of electrical or light energy on any of a variety of gases, solids or liquids. These are contained in a suitable confining enclosure called a resonator.

When an E.M. Wave, or light, of a particular wavelength passes through a suitable, highly excited (energized) material, it stimulates the emission of more E.M. Waves of the same corresponding wavelengths. This light or radiation is in step (*in phase*) with, and in the same direction, as the stimulating light source. The resulting high intensity (*AMPLIFIED*) and coherent product is the finished product of laser action. Through this medium, astounding scientific experiments may be performed.

Absorption, Spontaneous Emission and Stimulated Emission

The three basic interactions of photons with matter can be described with the aid of Fig. 6-1. Atoms or atomic aggre-

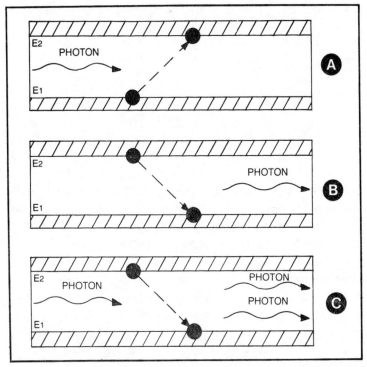

Fig. 6-1. Absorption (A), spontaneous emission (B) and stimulated emission (C).

gates can contain only certain discrete amounts of internal energy. They can exist only in certain discrete energy levels or states.

An atom will normally reside in the lowest energy state possible, called the *ground level*, unless it is given energy by some external means, or is stimulated. One way an atom can gain energy, or become excited, is by the absorption of a photon of light (Fig. 6-1A). However, absorption by an atom can take place only when the incoming photon has an energy exactly equal to an energy level separation E in that atom. Referring to Fig. 6-1 for instance, $E = E_2 - E_1$. If left to itself, an atom can lose energy by spontaneous emission, radiating a photon of energy E in any direction.

An excited atom can also be stimulated to emit a photon of energy E, if another photon of energy E strikes that atom. As a result, two photons will leave the atom, and most importantly, they will have the same wavelength (frequency), the same phase and the same direction. Thus, the stimulated emission process (Fig. 6-1C), is the basis of laser operation. It functions as a *coherent amplifier* on an atomic scale.

Population Inversion

In order for a laser to function, stimulated emission must predominate over absorption throughout the laser medium. The probability per unit time of the occurrence of each of these processes is the same, as was proven by Einstein in 1917. Therefore, for stimulated emission to predominate, more atoms must be put in the excited state than are left in the lower state. The distribution of atoms in the energy levels is then upside down, or inverted. This distribution is called *population inversion*, or the reverse of the normal ratio of excited to non-excited atoms in the mass population.

Mass Excitation

An inverted population can be produced or achieved by *energy pumping*. In insulating crystal lasers, this pumping is accomplished by the application of intense radiation with light of a higher frequency than that amplified. This is called *optical pumping* or mass excitation.

In semi-conductor junction lasers, the pumping is brought about by the use of electronic currents. It is thus called *electronic pumping*. In gas lasers, the pumping is achieved by electron-atom, or atom-atom collisions, and is thus referred to as *collision pumping*. In chemical lasers, the excited atoms or molecules of the medium are excited by means of chemical reactions. This is called *chemical pumping*. In some gas-dynamic lasers, the pumping is accomplished by means of the supersonic expansion of a gas which is classified as *gas expansion pumping*.

We have several possible means by which we may invert the population of a medium. Each means is dependent upon its own particular system. Once we have brought about an inverted population as such, the laser is then ready to amplify any light passing through it, or that which it may be triggered by. In order for us to use such a medium as a laser light source, or optical oscillator, the optical gain arising from the inverted population must be in excess of the loss in the resonator. These losses are caused by scattering, diffraction and reflector imperfections.

To reach this end, the optical path through the laser material is made long in one direction. Highly reflecting mirrors are arranged at each end so that they may repeatedly send or reflect the light back and forth many times along its length before releasing it from this resonating medium or

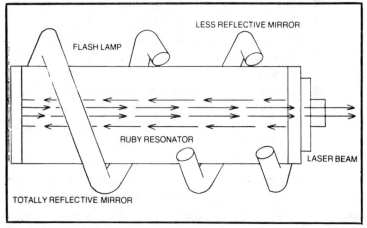

Fig. 6-2. An elementary ruby laser and its components.

enclosure cavity. The mirrors provide the necessary feed-back, similar to what is required in radio frequency oscillators. They convert the optical amplifier into an optical oscillator. Oscillation takes place only between the medium enclosed and is bounded by the mirrors. However, to affect the eventual release of the entrapped light, one end or mirror is made deliberately just a little less reflective. Or it may be made somewhat transparent to allow the laser beam to emerge from the resonator after having been sufficiently amplified.

One of the first perfected laser prototypes was the *ruby laser* (Fig. 6-2). The basic components of a pulsed ruby laser include:

- A light emitting medium, or substance
- A ruby crystal
- Two reflective mirrors that face each other and provide the confining barriers for the medium
- The resonating cavity, or resonator
- A controllable source of energy to excite the medium—a xenon-filled flash lamp (in a spiral form about the diameter of the ruby crystal)

Briefly, the pulsed ruby laser operates as follows:

- An electrical pulse or charge is set through the flash lamp, causing it to emit a burst of white light.
- A portion of this light is absorbed by the chrome atoms in the crystal.
- They in turn emit a red light in all directions.
- That portion of the red light that strikes the end mirrors is reflected back and forth many times.
- While in this process, it is being simultaneously amplified.
- Some of this red light passes completely through the one slightly less mirrored end of the resonator.
- It comes out as a highly monochromatic and highly directional coherent beam of collimated light energy.

The preceding description is very elementary and non-technical. It was given as more of a compass check—to see where we are along in our studies in preparation for our further clinical investigations into the applied real physical mechanisms of our system.

LASER CHARACTERISTICS

The shape of the resonator and the shortness of the wavelength of light make the laser emission look like a pencil-like beam (Fig. 6-3). A narrow beam is produced because many of the excited atoms are stimulated to emit in a specific direction, rather than in all random directions. This accounts for the extreme brightness of a laser. Laser light is also faithfully monochromatic. Its frequency spread is sometimes equal to only one trillionth of its generated frequency.

The use and functions of lasers depend on these mentioned characteristics. The intense brightness or power is the most important property for cutting, welding and drilling particularly difficult or otherwise impossible materials in industry. The monochromaticity of a laser is most important in spectroscopy. The shortness of the pulse is of paramount concern for high speed situations in the special fields of photography.

Fig. 6-3. Laser emissions take the form of pencil-like beams.

The combination of direction, high power and short pulse lengths make possible such long distance measurements as lunar ranging.

Amplification Control

Very instantaneous and high power outputs can be gained from a laser by using a technique called *Q-switching* of the resonator. In this technique, one of the resonator mirrors is made non-reflective during an interval of pumping of the laser medium. It then is suddenly made highly reflective. As a result of the switch, all the energy stored in the medium during the pumping interval is emitted in a powerful pulse of light, but it only lasts for about 10 billionths of a second.

The simplest way of *Q-switching* a resonator is to rotate one mirror very rapidly. Only during the brief time that it is lined up with the other resonator mirror (at the other end of the resonator) will laser emission occur. Another technique is to place in the resonator itself, a dilute solution of a dye that absorbs light at the laser frequency. The absorption of light by the dye initially prevents feedback. This causes an energy storage build-up in the laser medium. A point will finally be reached when the dye becomes saturated and can hold no more light. When this point is reached, feedback is restored and the laser emits an intense *Q-switched* pulse of light.

Mode Locking in Amplification Control

During Q-switching with a dye, further intensification of the output beam can be obtained by *mode locking*. In a mode locked laser there is a simultaneous oscillation of a number of closely spaced frequencies locked in time in a certain relationship to each other. The spectacular result is an even shorter pulse of but a few trillionths of a second in duration. During this brief instant however, the laser beam can reach tens of trillions of watts of power. This is more power than is being generated at any given instant than from all of the electric power stations in the world.

TYPES OF LASERS

There are a number of lasers available and each has a different use and purpose.

Insulating Solid Laser

The prototype laser used ruby as the light emitting medium. A ruby is a crystalline aluminum oxide containing a small amount of chrome. The chrome atoms are responsible for the red chroma of the emission. They are excited by optical pumping with green and violet light.

Initially the ruby laser could be operated only on a pulse, or intermittent basis because the pumping light had to be extremely intense. Later, it was operated at room temperature on a continuous basis rather than a pulse basis. Its major use now is as a high-power pulse laser. Operated in a Q-switch mode and followed by a second ruby laser used as an amplifier to intensify the pulse, a ruby laser can deliver several billion watts of power. These last only several billionths of a second.

Insulating Crystal Laser

The insulating crystal laser material typically consists of a hard, transparent crystal doped with a small amount of an element. This is often rare earth that has energy levels suitable for laser emission. A laser having such a crystal emits in the visible light, or near-infrared region. It often needs to be cooled far below room temperature to operate. Such lasers need intense optical pumping and they are usually operated only on a pulse basis to avoid overheating. Also, the output frequency of most insulating crystal lasers can be tuned only a small fraction of a percent by influences such as temperature or magnetic field.

Neodymium, a rare earth element, commonly is used as a dopant in various crystals or glasses because it has energy levels particularly advantageous for laser action with only modest optical pumping. The best crystalline host for neodymium has been yttrium aluminum garnet (YAG). The characteristic infrared emission from neodymium in this *host* at room temperature has been obtained on a continuous basis by pumping with a 100 watt incandescent lamp, or an array of light-emitting semi-conductor diodes or by using direct sunlight. The pumping by sunlight could someday provide us with an ideal possibility for spacecraft adaptation.

Glass doped with neodymium atoms is a useful laser material that radiates in the infrared region. Its properties are very similar to those of insulating crystal lasers. The neodymium doped glass obviates the difficult task of growing large crystals of high optical quality. The glass laser can operate continuously at room temperature and it has reached an efficiency of about 3% in pulsed operation. Neodymium doped glass lasers have produced the highest output power of any laser—tens of trillions of watts. This enormous power was obtained in Q-switched, mode-locked operations in pulses that lasted only a few trillionths of a second.

Semi-conductor Lasers

Crystal lasers can also be made from semi-conductors such as gallium arsenide (GaAs), lead telluride and others. Because these materials can carry an electronic current, electronic pumping of semi-conductor lasers is possible. A P-N junction like those used in transistors is formed in the semi-conductor crystal (Fig. 6-4). The junction is put in the forward bias with the positive voltage on the P side and the negative voltage on the N side. Electrons flow through the conduction band into the junction from the N type side. Holes flow through the valence band into the junction from the P type side. The conduction band is the upper energy level for the laser and the valence band is the lower energy level. Thus, an inverted population is established between the upper and lower energy levels. Then laser action occurs. Since the electron and hole flow are referred to as *injection*, these lasers are often referred to as *Injection Lasers*.

Injection Lasers

Injection lasers are efficient light sources. They are generally no larger than 1 millimeter (0.04 inch) in any dimension. For the most efficient operation, they must be cooled far below room temperature. For instance, 50% efficiency with a continuous output of several watts has been obtained by cooling a GaAs laser to −253 degrees C. (−423 degrees F.).

Most injection lasers can operate at very low temperature. However, an important exception is the GaAs laser. This laser can emit infrared light continuously at room temperature with an output of 0.02 watt and an efficiency of 7%. It

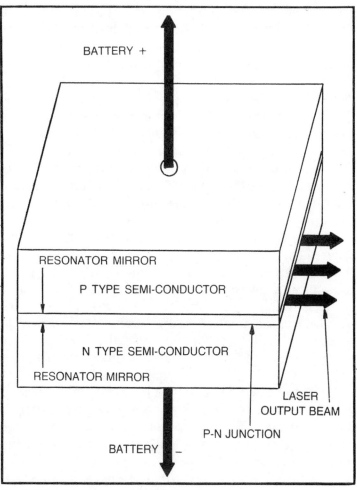

Fig. 6-4. A semi-conductor junction laser.

can be made in the form of a *HETROSTRUCTURE*. In the *hetrostructure,* a very narrow $P - N$ junction layer of GaAs is sandwiched between layers of a different semi-conductor— aluminum gallium arsenide. The properties of this semi-conductor confine the electrons and holes to the very narrow junction layer, leading to an inverted population at a lower input current. They also confine the laser light to this layer, thereby making the resonator very efficient.

Injection lasers can be fabricated to radiate at any wavelength between 0.64 micrometers and 32 micrometers.

This is done by alloying different pairs of semi-conductors, such as gallium arsenide and aluminum arsenide. Also, the radiating wavelength of an individual injection laser is sensitive to tuning by temperature, pressure and magnetic fields.

Gas Lasers

The first gas laser used a mixture of helium and neon that was *pumped* by an electrical discharge. Flat mirrors placed inside a long glass tube containing the gaseous mixture formed the resonator. Later, spherical mirrors placed outside the tube were found to be more convenient than mirrors placed inside the tube. Figure 6-5 illustrates such an arrangement employing outside mirrors.

The helium atoms are excited by collisions with electrons in the electrical discharge. An inverted population in the neon atoms is caused by the resonant transfer of energy during collisions with the excited helium atoms. This results in laser emission. This progress or sequence of collisions is a very effective pumping mechanism, but it is not widely applicable because of its dependence on a coincidence of particular energy levels. The helium and neon atoms provide such levels.

The original helium-neon laser is emitted at a wavelength of 1.15 micrometers in the near-infrared part of the spectrum. Radiation in the red part of the spectrum or at several other wavelengths of the infrared can also be obtained from this laser with only minor alterations.

Another laser that works on the atom-atom collisional pumping principle uses a mixture of nitrogen, carbon dioxide (CO_2) and helium. The CO_2 provides the laser emission. The CO_2 laser can continuously generate 16 kilowatts of power at a wavelength of 10.6 micrometers in the infrared region or several billion watts in a very short pulse.

Inverted populations can also be produced by electron-atom or electron-molecule collisions that occur in electrical discharges in gases such as neon, argon and carbon monoxide (CO), or in vapors such as mercury vapor and water vapor. Hundreds of gas lasers use this type of collisional pumping. In another type of collisional pumping, a high energy electron beam is used to irradiate a gas. This method has been used to

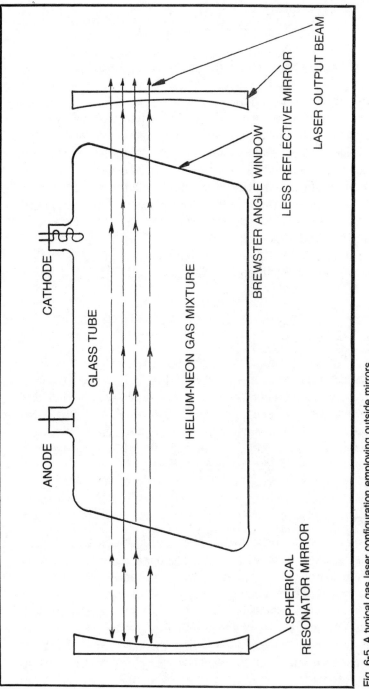

Fig. 6-5. A typical gas laser configuration employing outside mirrors.

95

obtain laser emissions in the vacuum ultraviolet region from hydrogen, xenon and krypton. In other gas lasers, a dissociation of molecules is obtained in the pumping process either by electron-molecule collisions or by the absorption of light energy. Oxygen, chlorine and methyl iodine are examples of gases used in these lasers.

Ionized atoms of a gas or vapor—that is, atoms with one or more electrons missing—may be used to produce a laser emission. A gas in this state is called a *plasma*. Argon, cadmium and neon have been used in this manner.

Sometimes gases are forced through the resonator as a supersonic velocity by using large engines. The supersonic velocity gases remove the excess heat generated in the electrical discharge used for the excitation. They thereby allow for the attainment of a continuous high power output. A laser that uses gases at supersonic velocities is called a *gas dynamic laser*.

Gas Dynamic Laser

The gas dynamic laser can also be used to produce gas-expansion pumping. If a uniformly hot mixture of nitrogen, CO_2 and water vapor is expanded at supersonic speed, the nitrogen pumps the upper laser energy level in CO_2 by collisions. This results in an inverted population and subsequent laser action. Continuous power outputs of 60 kilowatts have been achieved in this particular manner.

Chemical Lasers

In a chemical laser, a gas is created and pumped by means of a chemical reaction. The chemical pumping occurs through the release of energy in an *exothermic* chemical reaction. An example is the reaction of hydrogen and fluorine which produces hydrogen-fluoride molecules with an inverted population. That leads to laser action. Usually chemical lasers use the gas dynamic configuration so that the de-excited chemical reaction products and excess heat are quickly removed from the resonator.

Liquid Lasers

The most important class of liquid lasers uses a dilute solution of an organic dye in an organic solvent. Scores of dyes have proven useful. One of the most prominent is *Rhodamine 6G*. Laser emissions throughout the region from the near-ultraviolet (0.32 micrometer wavelength) to the near-infrared (1.2 micrometer wavelength) have been obtained by using various dyes. All are optically pumped, often with another laser. Several can be run continuously, while others can be mode-locked to produce pulses that have a duration of only a few trillionths of a second.

The most remarkable and useful property of a dye laser is the large and continuous *tunability* of its laser wavelength. In a single solution of slightly acidic 4-methylum belliferone, the laser emission has been tuned from a wavelength of 0.391 micrometer in the near-ultraviolet to 0.567 micrometer in the yellow-green region.

As you can see, there is a broad range of usable materials and substances available with lasing or masing capabilities. The various manners and techniques at our disposal affect such laser and maser emissions. You probably never knew that you carried a near-laser with you if you possess L.E.D. (light emitting diode) type digital wrist watch. Or that by observing a common neon display sign in a store window, you were observing the preliminary laser functions. These are to name just a few of the many daily functions that comprise our laser parts warehouse of ideas.

Chapter 7
Choosing A Laser

With such a variety of lasers available, it is important that you choose the appropriate one for your experimentation. However, as with almost any scientific endeavor, there are some very important safety precautions you must follow when experimenting with lasers.

BASIC CONSIDERATIONS

Your first consideration is whether to choose a polarized, or a randomly polarized laser. Linear polarization is especially important when the beam is to be measured after it reflects off of a surface, or is passed through an optical system. Linear polarization avoids the fluctuations in output power which otherwise occur when randomly polarized beams reflect off of surfaces that have polarized sensitive reflection or transmission characteristics. Use of a linearly polarized laser also permits control of the output intensity with a polarizer and permits the laser to be used with a variety of modulators.

Some versions of helium-neon lasers are available with linearly polarized output. The *HP* series lasers by the Hughes Aircraft Company have a linear polarization ratio in excess of 1000:1, while the *LF* and *LC* series lasers have a ratio of at least 500:1. If your intended use does not include the considerations discussed in the preceding paragraph, then you can probably realize a cost saving by selecting a randomly polarized version.

Your second consideration is choosing the proper power output level for your particular job. The output power requirement is determined by two factors: ultimate signal to noise ratio and power density incident upon the detector or recording media. These are factors which are often difficult to determine without experimentation in the user's particular application. As an example, in an alignment application, the detector may be the human eye and the noise source would be the ambient light. Experimentation would be required to determine the output power level necessary to overcome the worst ambient light conditions so that the laser beam would be visible to the human eye.

Your third consideration is that of *beam diameter*—spot size at a given distance. The width of a laser beam increases (diverges) hyperbolically with distance. Using two lenses to affect collimation, almost any laser beam can be shaped to the desired profile. However, it is best to select a laser which is close to the desired profile in the first place. Most lasers start with a waist at the outer face of the output mirror. Then they gradually diverge over a distance. It is easy to calculate the beam diameter, *d(cm)* at a distance *D(cm),* given the beam diameter d_0(cm), as it leaves the laser from this formula:

$$d = d_0 \left[1 + 6.492 \times 10^{-9} \frac{D^2}{d_0{}^4}\right]^{1/2}$$

This formula applies to helium-neon lasers at 632.8 nm only. All dimensions must be in centimeters (cm). For CO_2 lasers, which diverge more rapidly, the coefficient in the formula is 1.82×10^{-6}.

To maintain a small beam diameter far from the laser, a collimated beam expander should be used. This is a combination *double convex* and *plano concave* lens arrangement.) Hughes company offers a complete line of integral laser collimators and also separate collimators. One collimator, the Model 397OH, can be bolted onto some of the packaged helium-neon laser heads. The 397OH collimator expands the beam ten times to begin with, but then the beam diverges ten times more slowly. When the Model 397 OH collimator is used

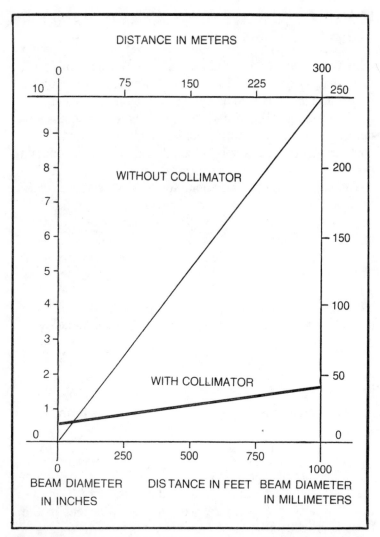

Fig. 7-1. Performance curve of a laser beam when used with a collimator and without a collimator.

with the Model 3076H laser it gives a significantly smaller beam beyond about 60 feet (20 meters) (Fig. 7-1).

SAFETY PRECAUTIONS

The following safety precautions should be used in working with, experimenting with or operating *ANY* lasers or other radiation emitting devices. They are recommended by The

Bureau of Radiological Health, Division of Compliance, Rockville, Maryland 20852.

- Never permit the eye to be in the direct path of the laser beam, no matter how low the power. The beam, whether infrared, visible or ultraviolet can cause serious burning of the optical retina of the eye.
- Avoid looking into the expected laser beam path when there is a possibility that the laser may be fired.
- Wear protective goggles (Fig. 7-2). Check laser specifications and available operating instructions for the maximum possible power and wavelength output. Make sure goggles are adequate for such expected radiation outputs.
- Even with goggles, do not permit the eye to directly intercept the direct path of the laser beam, the beam landing, alignment or projection.
- Keep specular reflectors out of the path of the laser beam.
- Avoid work environments with extremely low ambient illumination.
- Keep laser apparatus completely enclosed wherever possible. If operated without covering, then additional reflected radiation may be reflected.

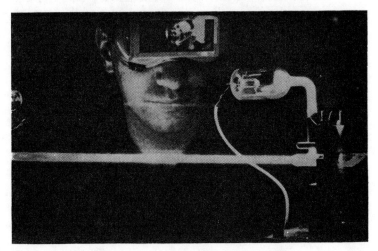

Fig. 7-2. Safety goggles are vital to any laser experimentation.

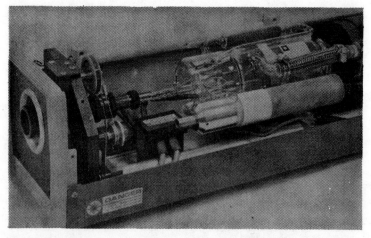

Fig. 7-3. Safety signs should be posted not only around the area of experimentation, but also on the laser itself. (Photo courtesy of Spectra-Physics.)

- The laser beam should be terminated at target by backstopping with carbon blocks.
- Never permit a laser to be left unattended while operating.
- A neon indicating lamp should always be made an integral part of every laser operating circuit. This is the case with all of the circuits given in this *Laser Experimenter's Handbook* since infrared and ultraviolet emissions are invisible to the operator and could exist without the knowledge of the experimenter.
- Signs should be posted (Fig. 7-3) properly identifying the operating area as an area of radiation or a laser test area. Care should be given that persons wearing or requiring heart pacemakers not be exposed to such radiations.

Chapter 8
Laser Projects

This chapter includes construction drawings and schematics for six easily constructed solid state, crystal and gas type lasers. Parts lists are also included.

The six lasers are:

- Project 1: L.E.D. pulse injection laser, 1–20 pps (pulses per second)
- Project 2: L.E.D. pulse injection laser, 50–2,000 pps
- Project 3: Helium-Neon (HNE) gas laser
- Project 4: Carbon dioxide (CO_2) gas laser
- Project 5: Ruby rod crystal type laser
- Project 6: General use gas type laser with a charging system, designed by the author

All injection (L.E.D.) lasers provided for in the plans are designed using the R.C.A. gallium-arsenide diodes. Substitutions are not recommended.

The laser in Project 4 is recommended for construction by the advanced experimenter only because of its inherent power.

Project 1: Pulsed Laser, L.E.D. Injection Type

Using R.C.A. # FLV-104 Gallium Arsenide Diode

This very basic and simplified L.E.D. laser (Fig. 8-1) should be constructed first in the order of experimental projects. It requires the minimum of parts and work for the experimenter. It will also provide the theory to advance the builder with confidence into more complex circuitry and components.

All of the necessary parts (Tables 8-1 and 8-2) may be ordered from the suppliers' list given in Appendix C.

The laser in Project 1 (Fig. 8-2) will emit a monochromatic red beam, capable of being visualized for approximately ½ mile with an adjustable pulse rate of from 1 to 20 pps (pulses per second).

This laser can provide the following uses:

- Optical alignment
- Long range intrusion alarm transmission
- Simulated weapons target practice
- Signalling devices
- Directional strobe light effects

LENS: 29 mm × 43 fl.
LENS MOUNT: 38 mm CAP
TUBE: 8″ LONG × 1.5″ O.D. BY—
.035″ WALL THICKNESS
LA-1: FLV-104 EMITTER DIODE
FOCAL LENGTH: TO BE ADJUSTED

Table 8-1. Parts List A for Pulsed Laser, L.E.D. Injection Type.

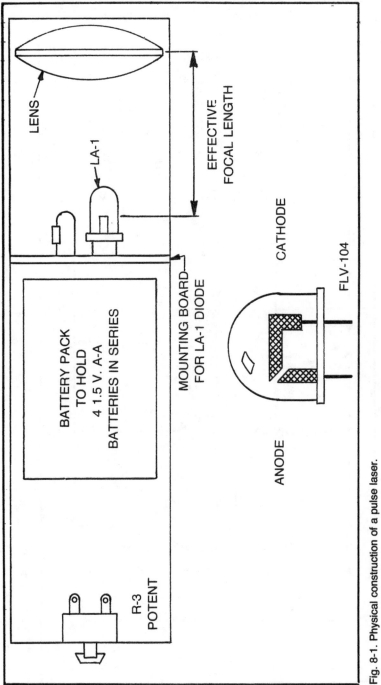

Fig. 8-1. Physical construction of a pulse laser.

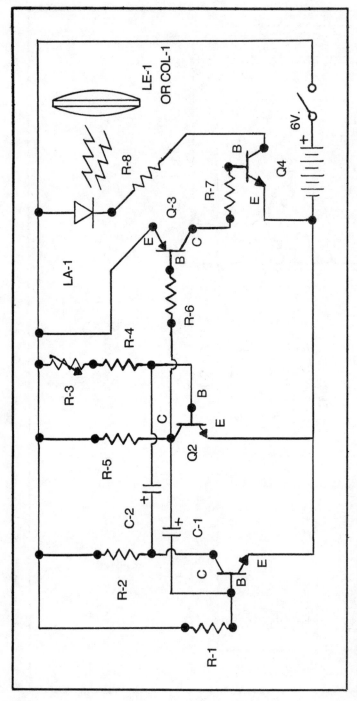

Fig. 8-2. Electrical schematic for a pulse laser.

LENS, 29 mm × 43 fl.
LA-1, FLV-104 EMITTER DIODE
Q-3, 2N2907 PNP
Q-1-2-4, 2N2222 NPN
R-1, 6.8K ¼ WATT
R-2 & R-5, 2.2K ¼ WATT
R-3, 500K. POT. & SW. COMB.
C-1 & C-2, 0.47/35V TANT CAP.
R-4, 15K. ¼ WATT
R-6, 6.8K ¼ WATT
R-7, 3.9K ¼ WATT
R-8, 27 OHM 1 WATT
BATT. SW S.P.S.T.

Table 8-2. Parts List B for Pulsed Laser, L.E.D. Injection Type.

Project 2: Pulsed Laser, L.E.D. Injection Type

Using S-6200 Series R.C.A. Gallium Arsenide Diode

The pulse rate of this laser (Figs. 8-3, 8-4, 8-5, 8-6 and 8-7) is adjustable from 50–2,000 pulses per second with a peak wattage of 1 to 20 watts. A 12 volt power pack or dry cell battery is required and a D.C. 12 volt battery charging system is included. See Tables 8-3 and 8-4 for additional necessary parts.

The diode and component housing is exactly the same as is diagrammed in Fig. 8-2. Thus, after the completion of the housing for Project 1, it may be used for Project 2 with no additional work involved.

Table 8-3. Electrical Parts List for Pulsed, L.E.D. Injection Type.

All resistors given in OHMS values
T-1: Square wave transformer, build to specifications given
D-1,2,3,4,5,6,7,8,12: IN2071 Silicon power diodes
D-9,10,11: IN914 silicon type diodes
Q-1,Q-2: D40D4 silicon power transistors
Q-3: UJT 2N2646 transistors
Q-4: 2N3439 Hi Volt NPN transistor
Q-5: 2N3583 Hi Volt NPN transistor
SCR-1: S.C.R. Motorola #2N4442 Rectifier
C-1: 100 mfd./25 volt vertical electric capacitor
C-2: 0.047 mfd./50 volt disc. electric capacitor
C-3: 0.3 mfd./400 volt metal. paper electrical capacitor
C-4: 5.0 mfd./150 volt electric CDE# NLW 5- 150 capacitor
R-1: 2.2K, ¼ watt
R-2: 220, 1.0 watt
R-3: 10, ½ watt
R-4: 500K, miniature potentiometer
R-5: 10K, ¼ watt
R-6: 100, ¼ watt
R-7: 33, ¼ watt
R-8: 1-2, ½ watt carbon type
R-9,R-12: 150K, ¼ watt
R-10: 1K, ½ watt
R-11: 330, 2 watt
NE-1: Neon indicating lamp
S-1: Switch, S.P.S.T.
S-2: Switch, S.P.D.T.
S-3: Normal open button switch
Battery 1 & Battery 3 consist of 4 - 1.5 volt D cells each, preferably NiCad., in series for 12 volts D.C. The optional battery charging unit required a 12 volt, ½ amp. transformer.

Table 8-4. Parts List for Pulsed Laser, L.E.D. Injection Type.

Tube: 10″ × 1.5″ O.D. × .035 wall aluminum tubing #6061-T6
Laser Tube: See Table 8-5 for power selection
Lens, LE-1: 35 mm × 31 mm Convex Edmonds #94-230
WR-1: Cable, shielded Beldon #8411
R-5, 120K: 2W carbon
R-2, A to J: 10 120K 2w resistors in series for 1.2 meg 20w.
C_1 & C_2: 0.15 mfd/4dv Sprague #430P
D_1 & D_2:V67 Varo
T_1: Transformer 4000V/.009 amp. Berkshire
NE-1; NEON indicator lamp

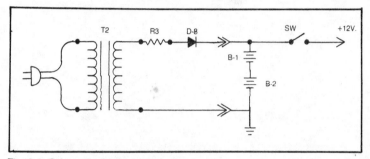

Fig. 8-3. Schematic for the battery charger.

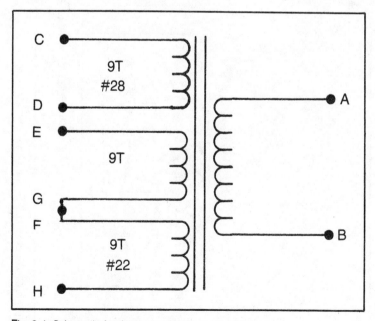

Fig. 8-4. Schematic for the square wave power transformer.

109

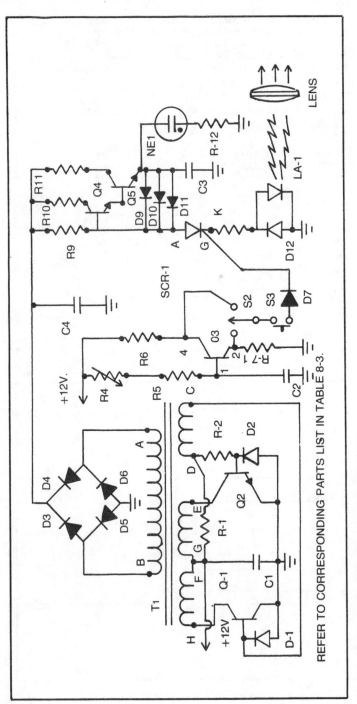

Fig. 8-5. Electrical schematic of a pulsed injection laser.

REFER TO CORRESPONDING PARTS LIST IN TABLE 8-3.

Table 8-5. L.E.D. Injector Diode Selection Chart.

L.A.-1:	POWER:	PEAK AMPERAGE:
All L.A.-1 Diodes are to be of the R.C.A. gallium arsenide type.		
#S62001	1-2 Watts	10 amps
#S62003	3-6 Watts	25 amps
#S62006	7-10 Watts	40 amps
#S62009	12-20 Watts	75 amps

The rate of pulse discharge is adjustable by C-2. Transformer T1 is affected by secondary coil wrapping. A change of one coil on the secondary equals a change of 1.33 volts. Refer to Tables 8-5 and 8-6.

Again, the rate of pps discharge on this laser is adjustable between 50 and 2000 pps. Use the darlington diode arrangement as shown in the schematic.

This L.E.D. type pulse laser is a more advanced system. It is of a good educational essence and should be attempted only after first building Project 1.

All safety precautions should be fully exercised while operating and testing this laser, especially since it is of higher output than Project 1.

The physical tube and housing for this laser assembly is also identical to that used in Project 1.

Table 8-6. Necessary Tube Information.

Tube No.	R_2	Power	Volts	Current	SeriesR.
IU05	1.2M	20W.	1100	4 ma	10-120K/2W
IU10	1.7M	50W.	1650	5 ma	25-68K/2W
IU20	1.7M	50W.	1650	5 ma.	25-68K/2W
IU30	1.5M	50W.	1800	5.5 ma.	25-62K/2W
IU50	1.5M	50W.	1800	5.5 ma.	25-62K/2W

R-3, R-4-Vari-Resistors; 0-150KOHM/2 Watt

Project 3: Helium-Neon Gas Laser

This project will provide the experimenter with a comprehensive insight into gas laser principles. It was selected and revised in its power aspects by the author for the purpose of safety and ease of construction.

This laser will emit a monochromatic beam of coherent light at a wavelength of 6328 nm. —corresponding to visible red.

Employing a ready-made and available *plucker* type gas filled spectrum tube model IO, this tube is a complete assembly. It can be ordered from Information-Unlimited (address in Appendix C). It includes internal mirrors of the spherical variety. The anode and cathode is factory sealed. No attempt should be made under any circumstances to open or otherwise tamper with the tube as provided.

This laser can provide the following uses:

■ Holography
■ Communications
■ Signalling
■ Alignment
■ Target practice for dry-firing weaponry
■ Any number of laser experiments

Depending upon the tube employed, a maximum power of up to 50 watts may be selected. All safety precautions should be exercised.

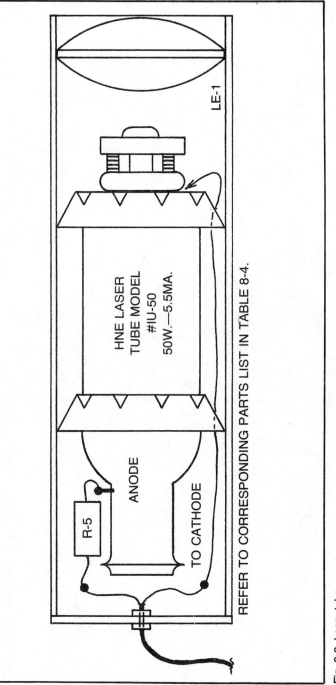

REFER TO CORRESPONDING PARTS LIST IN TABLE 8-4.

LE-1

HNE LASER
TUBE MODEL
#IU-50
50W.—5.5MA.

ANODE

R-5

TO CATHODE

Fig. 8-6. Laser tube.

113

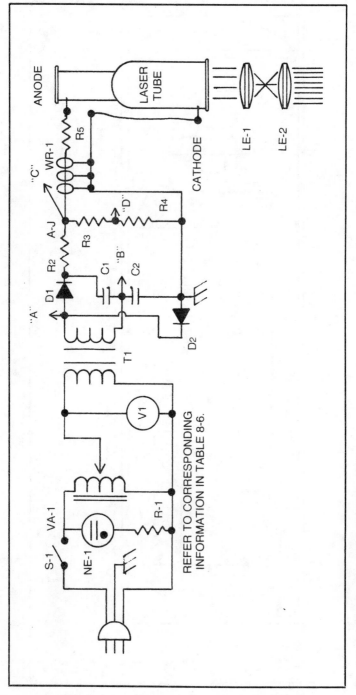

Fig. 8-7. Schematic with test points A,B,C, and D.

114

Project 4: CO_2 Gas Laser

The CO_2 gas laser produces an infrared beam of 10 microns wavelength. It has the power of 10 watts (peak) which may be used in the melting and burning of metals and other more serious demonstrations of high power laser effects.

This laser should be constructed and operated with the utmost attention being given to radiological safety. The emission from this beam could be extremely dangerous.

This project should be attempted only after the experimenter has gained knowledge and confidence by having first completed at least two of the preceding projects.

CONSTRUCTION NOTES

As with all gas lasers, the most important single component is the laser tube itself. The tube employed in this project can be purchased already assembled. It includes electrodes and a water jacket. However, the tube may be developed to any specifications the experimenter may wish.

The tube (Fig. 8-8) is ordered from Plasma-Scientific (address in Appendix C), cataloged as *MINilase 30*. It is made of Pyrex. The tube is 20 inches in overall length and has a laser bore of 15 mm. It is fabricated with an integrally blown water jacket for cooling purposes. If ordered from Plasma-Scientific, the Brewster Crystal Windows must be ordered separately and secured in place with epoxy glue. The electrodes, are however, included and factory sealed, leaving ample end wires for electrical connections.

A tube having any desired dimensions may be made by the experimenter. Note that the length of the tube and the effective volume of lasing gas between the cavity ends (Brewster Windows) has a direct affect on the power of the laser. With a length of 20 inches, as in this project, a maximum output of 10 watts continuous energy may be achieved. In industrial applications, the length of the tube may exceed 100 feet. An increase of 1 foot in tube length provides for an approximate increase of 10 watts output at the target. Even with the 20-inch tube, the output is approximately 10,000 times more powerful than the Helium-Neon laser.

The diameter of the tube is essentially controlled by the diameter of the Brewster Windows obtainable. Although the tube may be blown by the experimenter, it is suggested that he buy the tube ready-made. Glass blowing is an art in itself and this book is not designed to enable the reader to become an accomplished glass blower. However, if the reader has the nerve to attempt the job, the necessary information is provided in Fig. 8-13. The specifications are for a $20'' \times 15$ mm. tube.

The gas hook-up is shown in Fig. 8-10 where there are three steel bottles of CO_2, N_2 and He, respectively. They are connected to a manifold, with interposed needle values at the bottle fitting. I cannot stress the importance of using high quality valves for this project. The successful operation and maximum output of the laser is directly affected by the proper and ideal mixture of the three gases, therefore, the needle valves should have sensitive metering characteristics.

The gas manifold serves to receive the gases from their respective bottles and to provide capacity for their diffusion before reaching the tube. It must have a tap or fitting for connecting a manometer or accurate vacuum gauge. Get the best and most accurate manometer you can afford because the gas proportions and pressures specified must be achieved and maintained. The quality of the needle valves and the manometer is very important.

The vacuum pump may be a compressor from an old refrigerator. A good vacuum pump would otherwise be an expensive investment—about $100. An old or used refrigerator may be bought for about 20 dollars and the compressor may be stripped out of it in about 45 minutes. I personally have a Copeland Hermetic Compressor which I

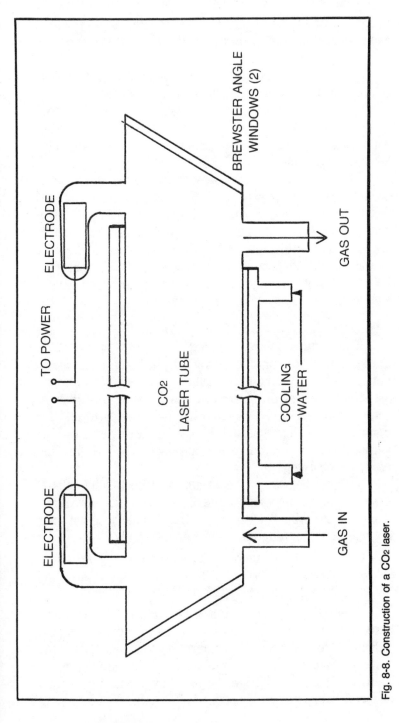

Fig. 8-8. Construction of a CO2 laser.

117

bought for $5.00 at a Flea Market several years ago. It works just fine and this type of compressor is very capable of pulling down a good vacuum when necessary.

Always keep the compressor lower than the laser assembly to prevent the oil which is in the compressor from accidently getting back into the laser tube and causing a difficult cleaning problem. A filter is also recommended to prevent this from happening. It's best to keep the compressor on the floor under the laser assembly.

The water jacket must be supplied with cooling water, preferably from a permanent tap. The discharge may be piped to a drain, unless you wish to reclaim it for some purpose. The water should be supplied at all times while the laser is operating because heat builds up quickly if operated dry. Make sure all water connections are tight and absolutely water proofed. Leakage of water could result in dangerous shocks due to the high voltages employed by this system.

Gas Charging

The mixture of gases is admitted to the system. With the bottles closed at their shut-off valves, the compressor is run until the lowest vacuum obtainable is reached. This may be assisted by heating the laser tube with hot water by a sponge. The system is allowed to stand for five minutes to test for tightness of all fittings. If the system can maintain a fair vacuum, then the system may be assumed to be leak-free and ready for charging.

The needle valves should not be used to shut off the system since the needles may be damaged by excessive tightening. After having been adjusted, they should be left at their setting, or the setting noted for future set-ups.

IMPORTANT: The ideal mixture or ratio of gases is as follows:

$CO_2 = 1$ part; $N_2 = 2.5$ parts; $He = 8$ parts

This ratio of gases is obtained by pressure adjustment. After the system is proven for tightness, crack open the helium valve and with the compressor running, adjust the valve to obtain a manometer reading of 4 millimeters (mm.) of mercury (Hg.). The nitrogen valve is next cracked open until an increase of pressure is obtained of 1.0 mm. We now have a

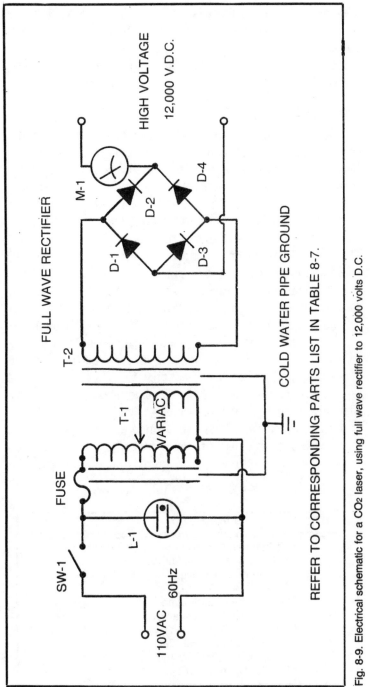

Fig. 8-9. Electrical schematic for a CO₂ laser, using full wave rectifier to 12,000 volts D.C.

119

pressure of 5.0 mm. Hg. Now we'll crack the CO_2 bottle valve and adjust for an additional increase of 0.5 mm. Hg. Thus there is now a total pressure of 5.5 mm. Hg. caused by the combined pressures of the three independent gases. The system is now gas charged.

Electrical Circuitry

Referring to Fig. 8-9, observe that the electronics of this system is pretty simple compared to the pulsed laser systems. The CO_2 system simply requires an electrical supply of direct current (D.C.) of approximately 12,000 volts with an amperage of at least 100 ma. A rectifier is provided to rectify the secondary 12,000 volts A.C. to a corresponding pulsed direct current. This is used to excite the gas mixture in the tube. This excitation in the gas laser context is appropriately termed creating a *plasma*.

A plasma is created when we ionize a gas by passing an electric current of high voltage through it. In this CO_2 laser, a plasma is created by passing a controlled current through the gas-filled laser tube. This causes the gaseous mixture to complete the electrical circuit between the electrodes of our pyrex tube and initiate lasing action.

The amount of current is controlled through the tube by adjusting the variable transformer—a variac (Fig. 8-9). By observing the current flow in (milliamperes) through the Milliameter M-1, and adjusting the mirrors and gas pressures, we can modulate the output to achieve maximum wattage.

Table 8-7. Electrical Parts List for CO_2 Laser.

QUANTITY	DESCRIPTION	PART NO.#
1	Milliameter; 0 to 100 ma.	
1	T-2, transformer, 12,000 V. 100 ma. neon sign	
1	Fuse, 110V. 10 amp	
1	L-1, neon indicator lamp	
1	T-1, transformer, variable, 0 to 120 V. 12 amps	5F743
4	D-1,D-2,D-3,D-4, rectifier diodes, 15,000 V. 100 ma.	
1	SW-1, switch S.P.S.T. 110 V. 10 amp.	

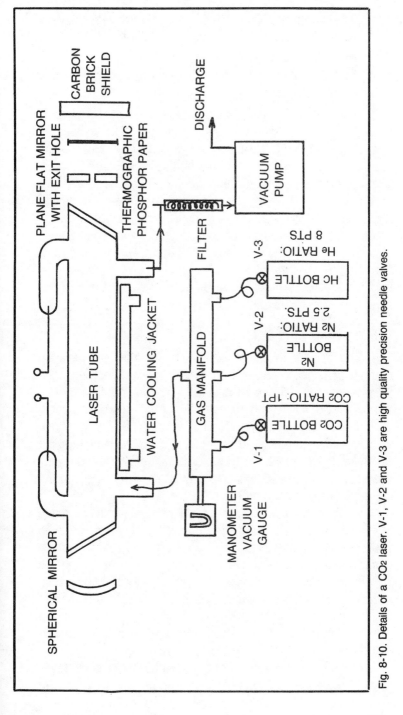

Fig. 8-10. Details of a CO2 laser. V-1, V-2 and V-3 are high quality precision needle valves.

121

Thus, the three variables for output modulation are:

- The current used for excitation
- The alignment of the mirrors
- The volume and ratio of the lasing material itself—the gas mixture

A neon indicator lamp has been included for a purpose. It is to provide an indication of system operation. When the system is energized, the indicator light will glow. This may be the only *observable* proof immediately available to prove that a beam is, or is not being emitted. The output beam of this laser is invisible, being in the infrared region—10 micron wavelengths. Thus, the system could be assumed to be off when there is no visible beam, yet the beam could cause damage. Do not omit this indicator lamp from the circuit.

No other electrical adjustments other than grounding the two transformers need to be made if the directions in the schematic are followed.

LASER STAND SUPPORT

The final and foremost variable for the efficient operation and performance of the system as a whole is affected by the ideal alignment of the mirrors or cavity ends of the laser. The mirrors serve as the end points of the optical resonant cavity.

It is imperative that the laser stand (Fig. 8-11) have rigidity and mechanical adjustment. This is so it can impart precision longitudinal and transverse movement for the purpose of achieving perfect mirror alignment with respect to the main axis of the laser tube.

Ideally, the base of the supporting stand should be made from a ¼ inch thick section of aluminum I beam, with a web width of not less than 4 inches and a flange width of anywhere from 4 to 6 inches. The laser supports and mirror supports should be made from aluminum or steel. They should be of at least ½ inch stock. The finished height can be left to the builder's convenience. However, it must accommodate a reasonable working height from the work bench it will ultimately be placed on.

A precise centerline must be considered before the drilling of any supports. This basic centerline will serve as the

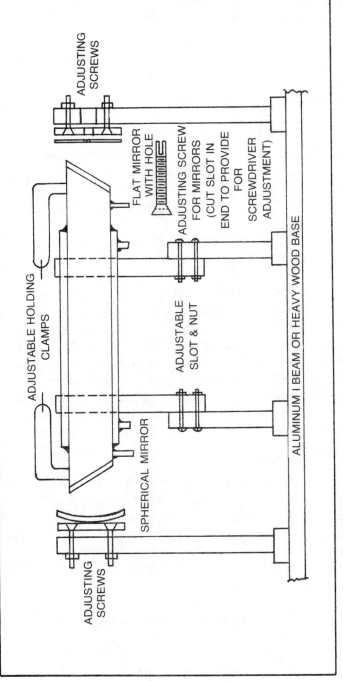

ADJUSTING SCREWS

FLAT MIRROR WITH HOLE

ADJUSTING SCREW FOR MIRRORS (CUT SLOT IN END TO PROVIDE FOR SCREWDRIVER ADJUSTMENT)

ADJUSTABLE HOLDING CLAMPS

ADJUSTABLE SLOT & NUT

SPHERICAL MIRROR

ADJUSTING SCREWS

ALUMINUM I BEAM OR HEAVY WOOD BASE

Fig. 8-11. Laser stand with mirror holders.

Mirrors: 1-Plano Type 1¼″ diameter
Exit aperature: ¼″ thick w/3 /32″
1—Concave: 1¼″ diameter ¼″ thick.
100″ Focal Length
Both Mirrors
To be Gold Coated
Order from
Plasma Scientific Min 30 MP,
Min 30 Mc respectively

Table 8-8. Parts List for Mirror Support Pads of a CO_2 Laser.

main longitudinal axis for all supports including the laser tube supports and the mirror supports. Thus, the more care that is given to laying out this initial centerline, the easier the system alignment will be when all of the components are assembled. It will also make the precision adjustment of the mirrors easier. This is vital because we're talking in terms of thousandths of an inch, which could affect the output beam.

At all times the mirrors (Fig. 8-12) must be at absolute right angles (90 degrees) to the main longitudinal axis of the laser tube. This is known as being symmetrically perpendicular to the main tube axis. The tube centerline must pass through the theoretical center of each mirror and be perfectly centered with the output mirror. This mirror has a 3/32 inch hole through which the beam will pass through. Although I have only shown the laser support legs to be vertically adjustable, the builder could also provide for the mirror support arms to have slots which would provide for the vertical adjustment.

Mirror Alignment

The mirrors are aligned in one of two ways. The first is to use a low power diode type laser as in Projects 1 and 2, or a high beam electric lamp. The second way is to use a surveyor's transit which will provide a very precise alignment. Industrial lasers are aligned by both methods. With a 100 foot long effective cavity between mirrors, absolute precision is mandatory.

Pieces of tin or cardboard are used, having a center hole of approximately ⅛ inch. Three such cards are spaced (Fig. 8-14) so the holes are aligned to provide a direct sighting through the effective optical length of the laser system from the exit end to the extreme reflective end.

The sighting instrument, whether another laser, a transit or a beam of collimated light, should be aimed at the entrance card from a distance of at least ten times the optical length of the laser. If the distance between mirrors were 2 feet, then we would use a distance of at least 20 feet for the sighting distance. This minimizes the projection angle.

The target beam, or sight point is shot at the center mark. It is permanently punched, or scribed at the appropriate center on the reflective mirror support facing toward the exit end of the axis. Thus, the line of sight (Fig. 8-14) shot from its

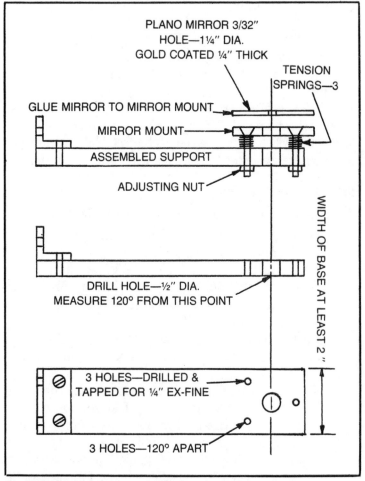

Fig. 8-12. Details of the mirror adjustment pads. The mirror support for the reflective end is the same as the exit except no hole is required.

source to the end support serves as a visual axis to center the openings created by the laser itself and the exit mirror support.

The mirrors must receive the maximum attention in their adjustment. They must be parallel to each other and perpendicular to the main laser axis. This requirement is achieved by the mirror adjustment screws on the mirror supports. Figure 8-11 illustrates the mirror adjustment screws and their geometry. The screws have slots cut in their ends to accommodate a screw driver blade. The mirror holder is attached to the mirror support by these screws. It has countersunk holes to receive the head of the screw and springs are provided between the support and the mirror holder to exert constant pressure away from the support. After once having set the mirrors, the nuts are tightened to lock the adjustment to that position.

The use of spherical mirrors makes the mirror adjustment easier than with flat or plano mode mirrors. In the CO_2 laser, we will be using a Hemispherical Mode of mirrors. These are represented by one spherical and one plano mirror.

Review of Construction

The following is a review of construction:

- The connecting leads from transformer T2 to the laser tube should be high voltage wires, such as automotive ignition cable. The voltage at this stage will be approximately 12,000 volts D.C. An accidental insulation breakdown at this point could be very hazardous.
- The filter shown in Fig. 8-10 can be a refrigeration type strainer obtainable from your local supply house.
- As shown in Fig. 8-10, the copper tubing connecting the gas bottles to the manifold should contain a *pig-tail* to facilitate slight movement of the bottles. Thus the laser assembly will not be disturbed. It will also further eliminate some transmittible vibrations which otherwise may affect the performance of the system as aligned.

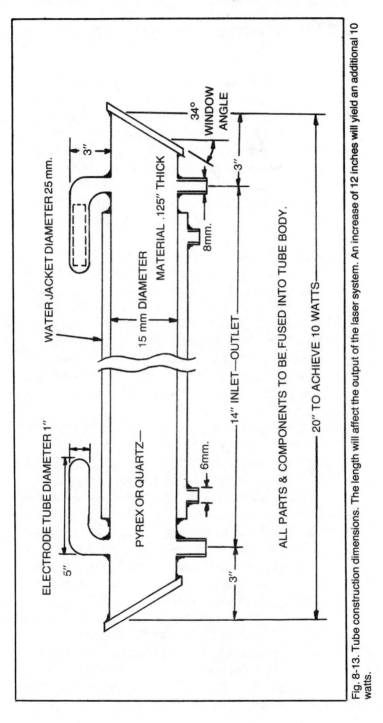

Fig. 8-13. Tube construction dimensions. The length will affect the output of the laser system. An increase of 12 inches will yield an additional 10 watts.

■ The discharged gas could be piped to a separate storage bottle for reuse if stringent precautions were to be observed. Maintain absolute purity of the mixed gas in respect to contamination and absorption of oil from the compressor. The discharged gas mixture is, after being mixed in the ideal ratios, capable of being directly admitted to the laser tube on recycling without the preliminary rationing. The gas at this time should be correctly rationed. Although additional time and energy will be required in planning a reclaiming set-up, it may pay for itself in the long run with big dividends because the three gases aren't cheap.

■ Before the laser is turned on, or at any time thereafter, make sure the aimed laser beam is properly back-stopped. This is done by placing a carbon brick or some other heat absorbing material behind the target. This prevents the beam from travelling indefinitely beyond the target and causing damage. A piece of waxed paper or a piece of thermographic phosphor paper may be used to indicate the presence or hit of an infrared beam. Place this in front of the brick back-stop.

■ In testing for leaks, first pump-down the tube by running the compressor until no more than 1 mm. of pressure exists. Now crack open the helium valve and admit enough He to cause a pressure of 15 mm. Turn up the current by adjusting T1 variac to produce 100 ma. and observe the color of the gas in the tube as it turns from a purplish color to an orange-pink glow. This orange-pink color indicates that no nitrogen (from the atmosphere) is present. It proves the system is gas tight and free of contaminates. This test is run with only the helium connected to the system. The possible valve leakage of N_2 would indicate a nitrogen contamination. This would lead us to believe that a leak to the atmosphere has occurred, when really the cause is from the nitrogen leaking at the valve. After an orange-pink glow has been witnessed with no trace of purple in it (purple

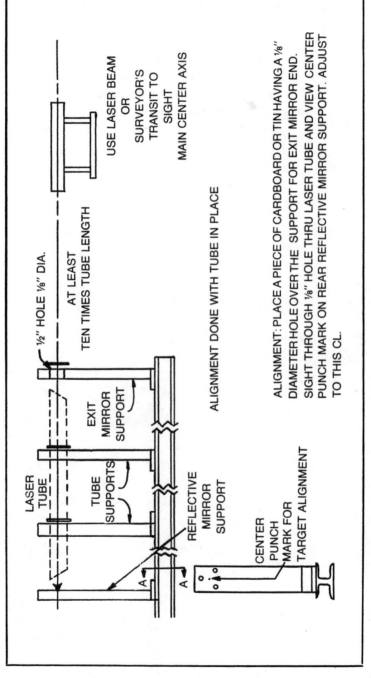

Fig. 8-14. System alignment showing functional axis.

½" HOLE ⅛" DIA.

AT LEAST TEN TIMES TUBE LENGTH

USE LASER BEAM OR SURVEYOR'S TRANSIT TO SIGHT MAIN CENTER AXIS

LASER TUBE

EXIT MIRROR SUPPORT

TUBE SUPPORTS

REFLECTIVE MIRROR SUPPORT

CENTER PUNCH MARK FOR TARGET ALIGNMENT

A

A

ALIGNMENT DONE WITH TUBE IN PLACE

ALIGNMENT: PLACE A PIECE OF CARDBOARD OR TIN HAVING A ⅛" DIAMETER HOLE OVER THE SUPPORT FOR EXIT MIRROR END. SIGHT THROUGH ⅛" HOLE THRU LASER TUBE AND VIEW CENTER PUNCH MARK ON REAR REFLECTIVE MIRROR SUPPORT. ADJUST TO THIS CL.

indicates the presence of N_2), we may then shut off the helium supply. While still running the compressor and when the pressure of the He gets down to about 1 mm. Hg., a whitish-grey glow should be witnessed of the plasma. If this color is realized, the system is absolutely proven to be leak-free. If not, check for leaks. Do not attempt to experiment until all leaks are corrected.

Project 5: Ruby Rod Laser

The ruby rod laser is a crystal type laser which will emit a laser beam of 6328 nm. (red). It is triggered by a pulse of 25,000 volts and powered by a 12 volt D.C. supply. This makes this system a portable unit for field projects and mobile uses. It is a fully transistorized model.

This laser can provide the following uses:

- Holography
- Signalling
- Weapons sighting
- Intrusion alarms
- Any project requiring a coherent and collimated source of light energy

Details are given for both circular and elliptical reflective systems (Figs. 8-15, 8-16 and 8-17). Since this is a pulsed laser, no cooling jacket is necessary. However, you must provide ventilation around the outside surface of the reflection chamber.

This laser is the prototype laser and should provide sound scientific experience in its construction and use.

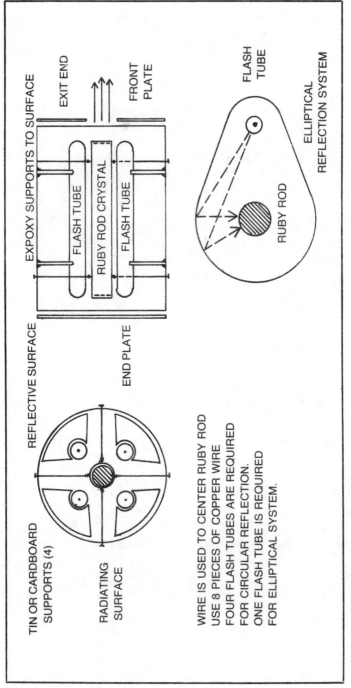

Fig. 8-15. Construction of ruby rod reflective system.

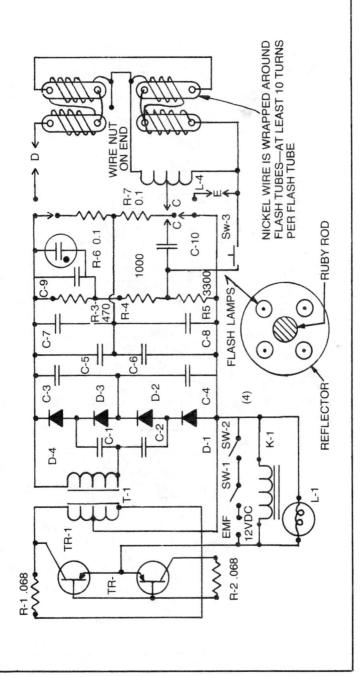

Fig. 8-16. Electrical schematic for ruby rod laser.

133

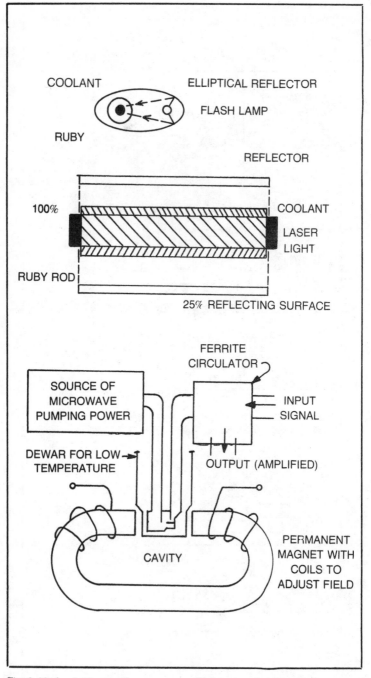

Fig. 8-17. A ruby laser with a maser amplifier system for 9000 MHZ.

Table 8-9.Electrical Parts List for Ruby Rod Laser.

TR-1, TR-2: Transistors, power 2 N 351 A, Texas Instrument
T-1: Transformer, center tap, 12/6 volt
T-2: Transformer, flash tube supply, Iron Core, 24 KV output
K-1: Relay, 12 volt D.C. 80 ma. D.P.D.T., 10 amp contacts
C-1,C-2,C-3,C-4: Capacitors, paper tubular, 1 mfd., 400 volt D.C.
C-5,C-6,C-7,C-8: Capacitors, electrolytic 2500 mfd., 350 volt D.C.
C-9: Capacitor, paper tubular, 0.05 mfd., 40 volt D.C.
C-10: Capacitor, paper tubular, 0.2 mfd., 400 volt D.C.
D-1,D-2,D-3, D-4: Diodes, 600 volt D.C., Fairchild IN4005
R-1,R-2: Resistors, power, wirewound type, 68 OHMS, 5½-watt
R-3: Resistor, carbon, 470 K OHMS, ½ watt
R-4: Resistor, carbon, 1 MegOHMS, ½ watt
R-5: Resistor, carbon, 3.3 MegOHMS, ½ watt
R-6,R-7: Resistors, power, 100 OHMS, 12 watt
SW-1: Switch, rocker, on-off, 10 amp. 125 V.A.C.
SW-2: Switch, S.P.S.T., (Optional for power panel interlock)
SW-3: Switch, push button
LASER RUBY ROD: Silvered, 4″ long, ½″ diameter, United Co.
FLASH LAMPS: 4 required, exflash 100, order form Plasma Scientific
L-2: Neon indicating lamp, NE-2 1/25 watt
L-1: Miniature lamp, bipin, 12 V.D.C., 60 ma.
Power supply: An automotive battery or standard D.C. converter may be used to provide a 12 volt direct current source.
Nickel wire: Approximately 4 feet required for flash tube trigger circuit use # 26, available from local hardware store.
High voltage wire: The wire to the flash tubes should be at least 10,000 volt ignition cable, automotive type. This voltage can reach lethal proportions should an insulation breakdown occur. Use stranded wire, not carbon filled type, size #10.
Reflector: This is nothing more than a No. #10 can having been polished on the inside to provide a reflective surface and darkened on the outside to assist in|dissipating the heat from the flash tubes. A better arrangement could be had by soldering heat fins on the outside to achieve faster heat dissipation.
Laser rod supporting wires: This is copper wire, single strand, #12 A.W.G. It is used to center and fix the ruby rod to the center most position of the reflector chamber. It is secured through holes in the outside of the reflector. The wires may then be soldered.

Project 6: General Use Gas Laser

With charging system for wide
experimentation projects of the builder's design

This system has been designed by the author to offer the experimenter a flexible set-up whereby he may select his own gaseous lasing material and power modes.

The author has included a schematic of his own charging system which represents the product of years of refinement and improvements.

The laser tube can be constructed by the builder or can be ordered by supplying the specifications provided.

If a continuous beam mode is selected, then a water coling jacket must be made a part of the assembly.

With the two compressors connected in series compound (Figs. 8-18, 8-19, 8-20 and 8-21), low vacuum phenomena may be experimented with in additional electron discharge projects.

Figure 8-18 illustrates the *"Six-A"* gas charging system valve arrangements and sequences. To purge the laser of

Table 8-10. Parts List for a Continuous Mode; Electrical Supply 10 Watt.

D-1,D-2,D-3,D-4: Diodes, 15,000 volt/100 ma.
T-1: Variac #5F743,0-120 volts, 12 amps.
T-2: Transformer, 12,000 volt/100 ma.
Milliameter: 0-100 ma.
L-1: neon lamp fuse: 10 amp.
SW-1: Switch, S.P.S.T.

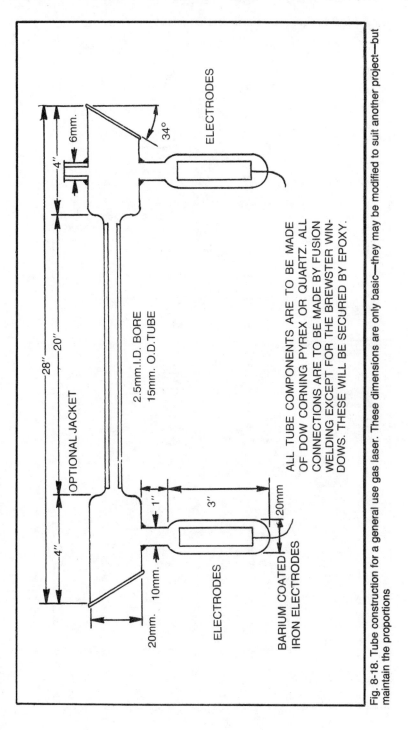

ELECTRODES

34°

6mm.

4"

28"

20"

OPTIONAL JACKET

2.5mm. I.D. BORE
15mm. O.D. TUBE

ALL TUBE COMPONENTS ARE TO BE MADE
OF DOW CORNING PYREX OR QUARTZ. ALL
CONNECTIONS ARE TO BE MADE BY FUSION
WELDING EXCEPT FOR THE BREWSTER WIN-
DOWS. THESE WILL BE SECURED BY EPOXY.

4"

20mm.

10mm.

ELECTRODES

1"

3"

20mm

BARIUM COATED
IRON ELECTRODES

ELECTRODES

Fig. 8-18. Tube construction for a general use gas laser. These dimensions are only basic—they may be modified to suit another project—but maintain the proportions

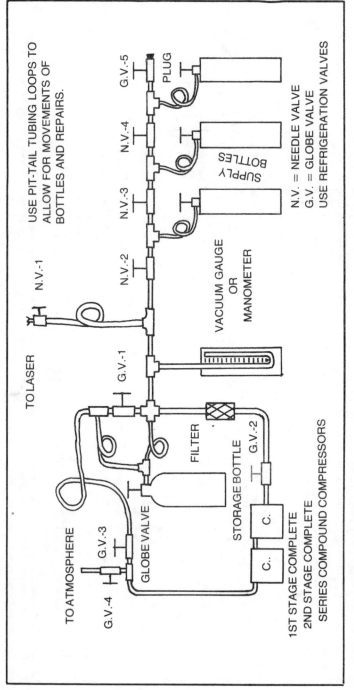

Fig. 8-19. Gas charging manifold for a general use gas laser.

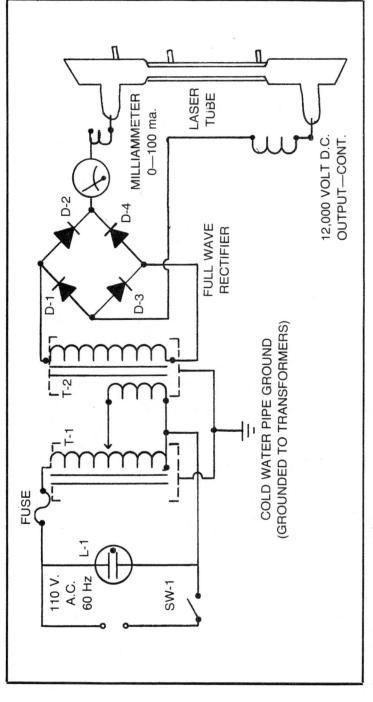

Fig. 8-20. Continuous mode; electrical supply 10 watt.

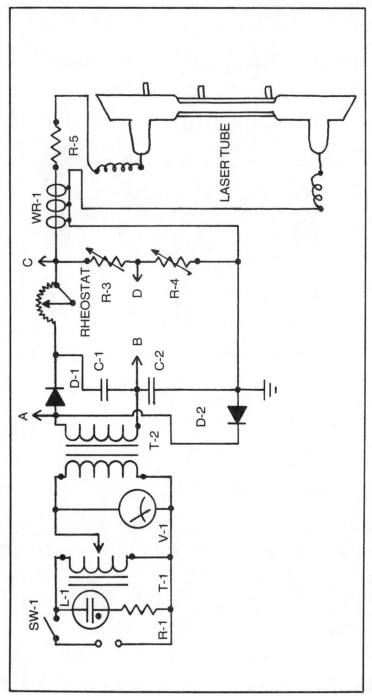

Fig. 8-21. Pulsed mode; electrical supply 50 watt.

Table 8-11. Parts List for a Pulsed Mode; Electrical Supply 50 Watt.

T-1: Variac Transformer 0-120 volt
T-2: Transformer 4000V./.009 amp.
Rheostat: 0-1.2 meg./20W.
C-1,C-2: 0.15 mfd./4000V.
D-1,D-2: diodes, V67 varo 5000 volt max.
A-B-C-D: test points for electrical adjustment
R-5: 120 K OHM-2W. Carbon type
V-1: voltmeter 0-120 volt: L-1, neon lamp
WR-1: shielded cable, Beldon #8411
R-3,R-4: vari-resistors 0-150 K OHM/2 watt

contaminated gases or air, open G.V.-2, G.V.-4, N.V.-1 and close N.2, 3, 4, G.V.-1 and 3. Run both compressors and discharge gas or contaminates into the atmosphere. To transfer the charge system to a storage bottle close G.V.-4 and G.V.-1, but open G.V.-2, G.V.-3 and any needle valves necessary to achieve the opening. Run both compressors until vacuum gauge indicates near vacuum. Valves G.V.-1 and G.V.-2 may be closed when charging from supply bottles, until pump-down or transfer again becomes necessary.

Chapter 9
Phasors

In chapter 2 on the Electromagnetic Wave Theory, we spoke of using phasors to arrive at effective or resultant values of E.M. Waves. We must bear in mind that for us to use phasors, or this application of mathematics in determining the wave's resultant value, that the waves must be of the exact same frequency. *All* of the waves, that is.

We may only resolve the resultant value for all waves of for example, red light, with a wavelength of 6328 nm. All other values for other wavelengths of different frequencies must be dealt with separately, according to their respective values. What we are specifically interested in is the effective power, or equivalent balance of the effect of many different E.M. Waves occurring simultaneously. These are the multitude of stimulated photon emissions within the laser or maser.

PARALLELOGRAM

These waves, although all of the same frequency and of identical wavelength, may, and quite often do, appear out of phase with each other. In other words, we may have some occurring 10 degrees apart in time, some 60 degrees apart in time and others occurring at another degree in time. These are either leading or lagging as circumstance may have them occur. Remember that all the waves we will be evaluating are

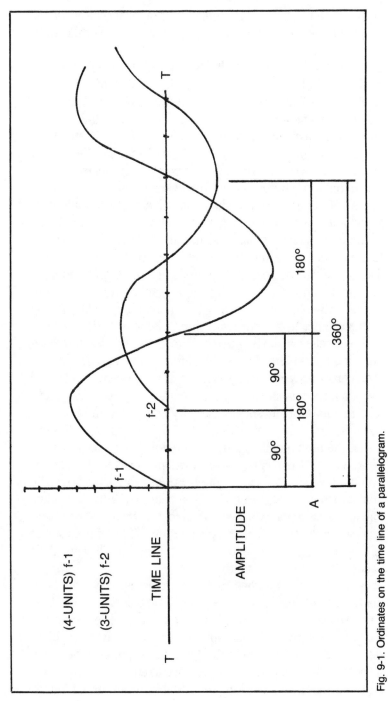

Fig. 9-1. Ordinates on the time line of a parallelogram.

143

of the same frequency. The only other differences the waves may have are the degrees apart in time from each other, their amplitudes from a given base line (Figs. 9-1, 9-2 and 9-3) and their amplitude in respect to each other. Actually what we will be doing is collecting the resultant effects of all of the waves' total effects. This product is called the *resultant* of the individual components. Thus, in Figs. 9-1, 9-2 and 9-3, the arrows, or measured scalar arrows f-1, f-2, f-3, f-4, etc.,) are the component sides of an exact parallelogram. For every instance, we will construct them on paper with exactly proportioned units of length. Thus if we were to have two components with lengths of 3 inches and 4 inches, we would construct a geometrical parallelogram with the sides of the box 3 inches and 4 inches. A diagonal measured across the interior of this box, or parallelogram, would measure exactly 5 inches. The phasor for this particular parallelogram with components of 3 units and 4 units would yield a resultant or phasor of 5 units. Although this is not quite an outright addition or a multiplication, it is a sort of compromise.

We select a starting point which is usually the first component off of the *time line* in a counterclockwise direction. We then construct a parallelogram for this component using it and the component adjacent to it (in a counterclockwise direction). The resultant or phasor obtained by computing the first two components then serves as a side, or a component to construct the next parallelogram for the next succeeding component. This process is repeated, working in a clockwise direction until we return back to the *time line* where we originally started from. In place of all of the representative individual values or components, we resolve for them, a resultant phasor equivalence. This represents the effect of the total of all the individual components if they were to all have their chance to act independently.

In Fig. 9-1, we construct our ordinates. Have the horizontal line represent the time. This time line is marked off in time degrees—360 units units equalling one complete wavelength, or cycle. Comparatively, we may be able to determine, by reference to this line, the relationship between adjacent waves in respect to their occurrences to each other. Not only can this proximate relationship, whether *leading* or

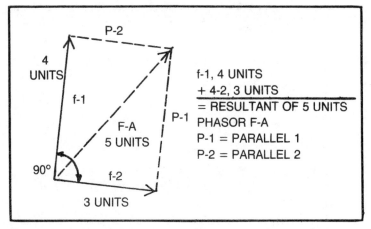

Fig. 9-2. Resultant of a phasor parallelogram.

lagging behind the first wave to be evaluated, be constructed, but it's quantitative aspect can be determined in terms of the number of degrees apart these waves are. Thus a wave may be said to lead its neighbor by 45 degrees. Or conversely, we can describe this by saying that the one wave lags the other by 45 degrees, depending upon which wave we wish to refer to as a starting point or reference point.

By laying-off the observed or reported amplitudes of each component wave to its respective scalar magnitude (Fig. 9-2) and laying these components out in a geometric arrangement (such as would equal their exact time line relationship in degrees lead or lag), we will have the two sides (f-1 and f-2) of a parallelogram. These two sides (f-1 and f-2) represent the laid-off amplitudes and relative angle of displacement as observed in the wave diagram in Fig. 9-1. All that is now needed is to use a parallel ruler and draw sides P-1 and P-2 exactly parallel to components f-1 and f-2. Thus we now have a parallelogram having an angle of 90 degrees, or whatever angle of time displacement the problem may dictate.

Draw a diagonal line between the original apex of the 90 degree angle. Connect this angle with the point of angular intersection made by the parallel lines P-1 and P-2. Now in the center of the parallelogram there is a diagonal line which is the actual phasor or vectorial resultant of the two sided components, f-1 and f-2. You now know how to construct a *phasor*

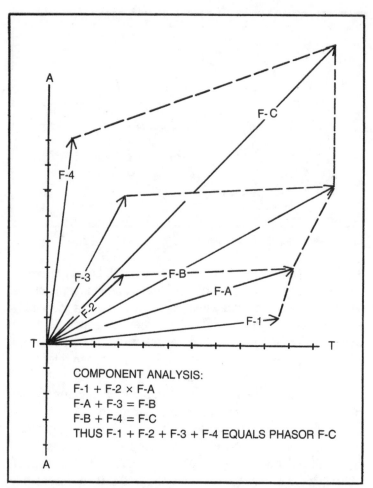

Fig. 9-3. Component analysis of a phasor diagram. The vertical sum of all waves or values equals the equivalent of one wave having the effective value of F-C.

parallelogram. By employing this vector system of computing components of complex waves or quantities, you may be able to achieve from any number of such complex, but qualitatively similar scalar values, one resultant or phasor sum. This phasor system can be used to derive a resultant phasor from any number of components. The components can be either all positive, or positive and negative components. Handle all above-the-line positive values and all below-the-line values as one component each of a parallelogram. The resultant will be an effective difference. Attention should be paid as to whether the components are picked-up in a counterclockwise (positive) or clockwise (negative) manner.

COHERENCY

The coherency of laser light, meaning the *in phase* characteristic of laser light, permits us to apply such a *phasor analysis* to determine the effectiveness or resultant power of such an E.M. Wave. Ordinary incoherent light, by virtue of its multiplicity of wavelengths would make this application impossible. The main body of laser emission is coherent and of the same frequency, differing only slightly in its exact phase. It keeps pretty near to the absolute 0 degree phase as the inherent method of generation. Stimulated photon emission assures all photon emissions to have occurrence at the same frequency.

Figure 9-3 illustrates the vectorial system in which a polyphase analysis is made of four separate values. The diagonal (resultant) of the first completed diagram serves as one side of the next parallelogram to be constructed. This progression is duplicated for each and every component to be vectorially collected.

Chapter 10
The Laser: Yesterday, Today and Tomorrow

In 1958, Dr. Charles H. Townes and Dr. Arthur L. Schawlow had advanced the propositions of how light may be made coherent by the amplification of stimulated emission of radiation.

HISTORY

These two scientists took principles one step further and in place of microwave propagation, had the end product appear as light energy. The maser was invented in 1951 by Dr. Townes. This invention brought him the coveted Nobel Prize in 1964. The maser is a device for the amplification of very feeble microwaves and for the transmission of the waves in a more perfect manner.

In 1960, we find that Dr. Theodore H. Maiman, who was then working at the Hughes Aircraft Company in Malibu, California had succeeded in actually applying the theories. He was rewarded in achieving from a synthetic ruby, a very minute beam of red, coherent light. This was the first such beam of laser light to be propagated by the laser process.

From these humble and seemingly infantile roots we have now a marvel which is surpassed by no other basic invention since the wheel. During the printing of this manuscript, even newer uses were found for the laser to perform. New frontiers

Fig. 10-1. New frontiers for the laser are constantly under investigation.

149

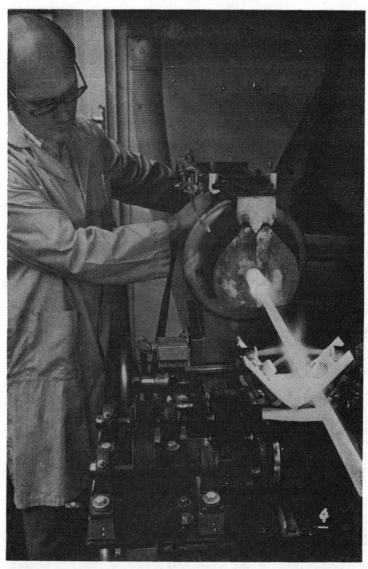

Fig. 10-2. A scientist evaluates this light beam.

for this electronic pioneer to hew for the satisfaction of mankind are constantly under experimentation (Figs. 10-1, 10-2, 10-3, 10-4 and 10-5).

The first, or prototype laser was but a small man-made ruby rod, about a ½ inch in diameter and approximately 2 inches long. It was silvered on each end to serve as a resonant

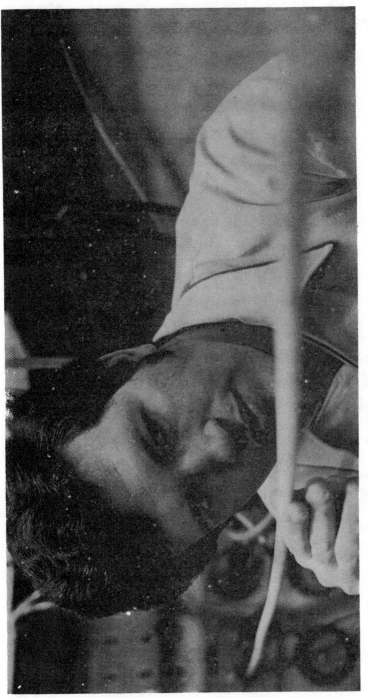

Fig. 10-3. Every aspect of the laser system is investigated by scientists.

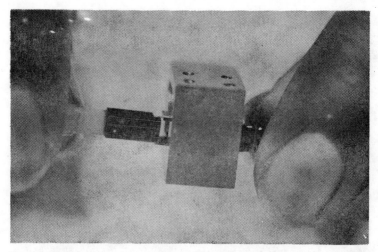

Fig. 10-4. Minute lasers are capable of tremendous powers.

chamber. One end was made slightly less reflective than the other. Surrounding this cavity was placed a wrap-around flash tube containing Xenon gas—a flash tube very similar to those used in high speed photography. When this flash tube was made to flash, not all, but a substantial amount of the light was absorbed into the ruby rod. The absorbed energy then actuated the chromium atoms which were added to the ruby material—giving the ruby amplification characteristics. This addition was called doping. The chromium atoms, once having been energized by the externally pumped light, then became elevated to a higher energy level, sort of like being kicked by an external body. It was known in advance that the chromium atoms could not stay at this unnaturally higher energy value very long—about a millionth of a second to be exact. Then the chromium atoms would drop back (decay) and give up their shortly held energy to the surrounding ruby atoms. This activates them and causes lasing.

The ruby atoms now begin to lase, which as we have covered, is where the photons of energy are surging back and forth, from end to end between the confines of the silvered resonator ends. They do this until enough of the atoms have developed the required threshold energy to cross the threshold of the mirror resistance. They break through and emission occurs.

If you think the prototype laser was small for its ability, then consider this—Bell Laboratories has already manufactured a fully functional laser which is smaller than a grain of salt (Fig. 5-12). Of course its power is accordingly diminished by its size, but it is a functional laser nevertheless. On the far extreme, gas lasers (Fig. 10-6) have been built to lengths in

Fig. 10-5. New discoveries are made every day due to the constant experimentation with the laser systems.

Fig. 10-6. A researcher operates a gas flow laser which uses a mixture of carbon dioxide and nitrogen.

excess of two hundred feet. The maximum output power of these gas lasers is directly proportional to their length between mirror ends, which is to say their resonance length. Thus, we have quite a variety in sizes and powers. They range from the carbon dioxide laser which can vaporize diamonds and all substances known to man, to the gentle surgical laser now being used to perform delicate eye surgery while the patient is fully awake.

The power ranges of lasers begin at the minuscule scale, which even a mosquito couldn't feel, to the model made by Westinghouse, where without even focusing, the output is in excess of 750 trillion watts of power. This is analogous to all of the water going across Niagara Falls being shot in one instant squirt, through a water pistol!

PRESENT APPLICATIONS

The doors of industry opened by the laser are countless (Fig. 10-7). And more are being opened each day. Metals and elements which were once thought to be beyond destruction are literally sent up in a cloud of bluish-green vapor at the

Fig. 10-7. Industrial production laser. (Photo courtesy of Coherent Laser Division.)

laser's concentrated thermal touch. Let's have a look at some of the other uses of the laser in today's technology.

Holography

Holo (whole), the practice of taking or projecting lifelike three dimensional pictures of an object was born as a result of the laser. With a *hologram,* we can visualize the entire picture in depth—even around the corners and opposite sides of objects.

The reason that the practice of holography was only recently developed is that the art itself depends upon the division of light rays in a simultaneous and exact manner. Actually, the image the observer sees is the result of the reformation of the reflected image at a distance from the lighted object. These reflected rays are then reconstructed at the point of junction wherever the mirrors are aimed.

Although not quite holography, the laser is shrewdly employed in museums to detect forgeries of valuable portraits and sculptures. In this application the laser produces a tiny controlled beam of light which vaporizes a minute particle of the item being tested. This vaporized gas is then viewed by a spectrograph and the elemental properties can be scrutinized. The laser for this operation is known as a micro-probe.

Medicine

In optical surgery, a controlled laser beam is used to perform surgery that only a few years ago was considered almost impossible. The laser device used by optometric surgeons is called a *photocoagulator.* It can literally weld the retina back in place for those thousands of people yearly who otherwise would suffer blindness from the separation of the retina in the eye. This operation is a very safe and successful one due to the laser. This operation can permit a patient to have the retina welded, then drive himself home from the hospital in a matter of an hour. This is about the same period of inconvenience as having a tooth pulled. The photocoagulator makes use of the lens of the patient's own eye to focalize the light energy to just enough intensity to fuse the delicate membrane of the retina.

The laser also finds its place in dentistry, where the teeth, after losing their hard glossy outer shell, are irradiated by the laser. It has been conclusively tested that teeth treated in this manner definitely are affected in the retardation of otherwise decay.

FUTURE

The *biolaser* is another application of the fundamental laser. Its characteristic ability directs pinpoint thermal energy in absolutely controlled dosages. It is utilized in destroying cancerous cells on the microscopic level. The beam used is kept to a diameter not exceeding forty-millionths of an inch in diameter. This art has now been perfected to such an exact degree that even chromosomes may be split at their nucleus. Bloodless surgery is another factor in favor of using lasers in surgery. Because of the concentrated heat at the very point of cutting, the focalized beam cauterizes the tissue as it incises, thus it closes off all arteries and capillaries as it progresses through the incision. At this writing, biolasers are successfully being used in the removal of tumors and skin scars with very little notice left to the area of removal. This is very desirable in cosmetic surgery where the removal of birthmarks might otherwise leave a scar worse than the original mark. This treatment has also proved to be very effective in the removal of tattoos, which were once assumed to be permanent.

You ask how it feels to have a laser beam touch your skin? Not any more painful than by just having a very small drop of melted wax drop on your skin. It's even better because there is no after sting or soreness since it affects such a very localized portion of the area being treated.

Radio Physics

It seems that the entire science of radio, television and communications had just breathlessly awaited the advent of the laser. The laser (Fig. 10-8) has opened new horizons in telecommunications that are just unbelievable. A few examples include telestar reception, where we can hear the bursts of light and radiation waves from stars which emitted them over a million years ago. We can also hear radio waves that were created long before even dinosaurs were roaming the earth. The heavens are so full of *star talk* that the maser has to be discretely tuned to the particular frequency sought. The bleed-over is enormous with all of the stars talking (creating E.M. Waves) at the same time. Who knows, perhaps very shortly we'll receive and be able to transmit in return, some very definite intelligence to one of these stars so far out from our solar system that they cannot even be seen, not even with the most powerful of all telescopes. But we may be able to hear them all the same.

By using a laser beam for transmitting communications in place of the conventional copper wire, we may transmit millions of telephone conversations simultaneously on the same beam. Thousands more television programs may be transmitted due to the inherent stability of the laser bandwidth of emission than can be provided by the carrier wave of say 10×14 Hz of a beam.

The laser finds unequalled use in radar detection. At this very instant, military lasers are scanning the hemisphere in alertness for anything which appears to be suspicious or lethal to our protection and security. Lasers are used as blind vision detectors for directing planes and missiles.

Experiments have even been conducted for some time to develop a system where the visually blind may be equipped with a laser transceiver system which would enable them to detect the presence of objects in their paths and indicate what direction the user must take to avoid contact with the object. It is hoped that the system may also be able to provide the user with a one direction outline or scan of the object detected. This would, in some respects, restore a type of vision to the blind. When perfected, the system should be able to at least perform everything that its canine equivalent does except bark.

Electrical power may be transmitted by laser beams, but only with an appreciable loss of efficiency. When this problem is eventually overcome, we can expect to witness the disappearance of all the electric poles and transmission lines from our country side, and have unnoticeable placements of antennae for transmission and relaying of laser beam conducting paths. We should also be ready for the news that scientists have developed the laser output to be adequate enough to

Fig. 10-8. A semi-conductor laser magnified 200 times, rests on a copper heat sink. Atop the laser is a wire that carries current to the device. Light exits through the cleaved edges. (Photo courtesy of Bell Laboratories.)

destroy missiles in our skies other than our own. This date of perfection will introduce us then to the age of absolute electronic superiority. The nation which will have the most sophisticated and capable laser systems will dominate the world, and ultimately our known universe. Let us then strive to have *our* nation so readied!

Industrial

The laser finds its greatest applications in machining, welding and cutting operations. What once were considered to be unusable materials because of their resistance to machining are now rendered almost pliable to the intensity and purity of the focalized laser beam. There is no substance known to man which cannot be dealt with by the laser. Pieces as small as radio tube filaments and components may be precision welded *while still inside of their vacuum enclosure.* This is ac-

Fig. 10-9. A dye laser pumped by a Coherent ion laser. (Photo courtesy of Coherent Laser Division.)

complished by the beam passing through the glass walls of the tube and focalizing at a convergence point inside the tube. Thus expensive electronic tubes may be salvaged which otherwise would require replacement. It is no problem at all to fusion weld a cathode inside a sealed vacuum tube for a laser. This is done without even removing the tube from the set and is being done all of the time.

With system guidance for machining processes, precision can be assured to the millionths of an inch on lathe operations. It can be as close as ½ inch on tunnel excavation work where tunnels several miles long may be begun at both sides of a mountain. The meeting point will be as close as either plus or minus ½ inch. *Interferometers* have been used to detect changes in length, such as in bridge spans to a billionth of an inch!

Fig. 10-10. Metrologic neon laser. (Photo courtesy of Metrologic Division.)

The conventional *gyroscope* used for directional stability has now met competition that the mechanical gyroscope can't compete with. The new laser gyroscopes are communicating to us about changes in the earth's orbital path that we would have never realized before. The mechanical gyroscope is dependent upon the earth's gravitational influence where the laser is electronic and non-affected by gravity.

Formerly, the process of drilling the small holes in diamonds used for wire drawing dies was a two day job, tedious and expensive. Now the job is accomplished in a few minutes by using a pulse of laser energy.

The precision of the laser in manufacturing (Fig. 10-9) is best related by its use in *trimming* carbon type resistors. Minute quantities must be removed to correct the resistors

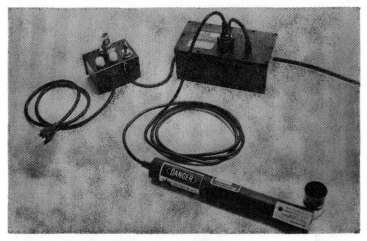

Fig. 10-11. Coherent CR-135 helium-neon laser. (Photo courtesy of Coherent Laser Division.)

Fig. 10-12. This researcher communicates via the laser communicator receiver and the Metrologic neon laser. (Photo courtesy of Coherent Laser Division.)

ohm values. This is an amount equal to one-billionth of a gram. This amount is so infinitesimal that its removal can only be proven by electrical units so small in value that another laser must be used to confirm it.

Since the appearance of the very first ruby rod type of crystal laser in 1960, science has introduced many additional lasers (Figs. 10-10, 10-11, 10-12 and 10-13) deriving their lasing ability from diverse properties. Essentially any substance can be made to lase under the proper influences. The possibilities are limitless as far as the future is concerned. The only comment that could safely be made is there is no definitive end point to the future of this sophisticated electronic marvel. It is so simple, but yet so profound.

The laser was here since the first crystal was formed on the face of this earth. Man had unknowingly assembled the gas laser as early as the late 1800's when gas filled geiger tubes were first experimented with. How close man was then to creating the catalyst that would someday affect the lives of us

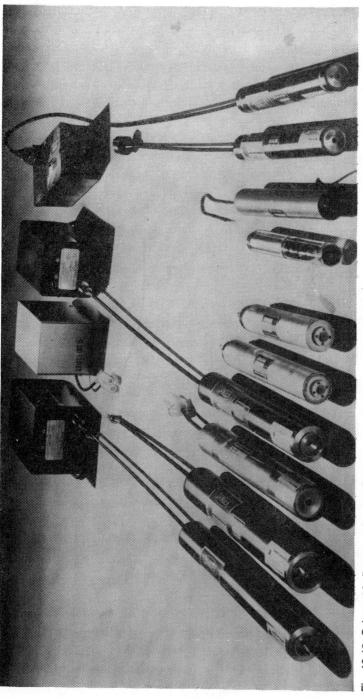

Fig. 10-13. Coherent helium–neon lasers. (Photo courtesy of Coherent Laser Division.)

163

all! Only man's cerebral inertia kept him from reaching out to the unknown, to dream things that are not, and ask ... why not? We must spare no energies now in the continued development and research of this natural instrument of the universe. Pray our advancement will lead us not to destruction, but rather to the civilization of our dreams.

Chapter 11
Advanced Technical Data

One of the newest lasing materials introduced by R.C.A. is the mixed crystal GaAsSb. It is a mixture of the two crystals gallium arsenide and gallium antimonide. Mixed crystals in general have the special property that the wavelength of light they emit may be tuned by adjusting their composition. For example, by increasing the amount of antimony in GaAsSb, the light emission can be continuously tuned to any wavelength from 0.9 to 1.2 micrometers.

MIXED CRYSTALS

Mixed crystals are not new inventions. Gallium arsenide phosphide has been used in commercial red light emitting diodes (L.E.D.). Aluminum gallium arsenide is an integral part of the GaAs lasers now being used in the Atlanta Lightwave Communication System. However, it must not be assumed that mixed crystals can be formed for any given set of elements. Whether or not a solid mixed crystal of a given composition can be formed depends on the forces between the particular atoms in the crystal and on the respective laws of thermodynamics. For GaAsSb, the limit for the GaSb content is more than 25%. This makes it possible to have lasers with wavelengths as long as 1.2 micrometers.

The new laser consists of a very thin layer of GaAsSb sandwiched between P-type and N-type AlGaAs wafers. These two wafers or layers inject electrons and holes into the gallium arsenide antimonide active layer and provide barriers which confine the electrons and holes in that respective layer. The confinement of both electric charges and light is a key feature of this layered *double heterostructure*. This makes possible the efficient operation of the laser.

The multiple layers for the lasers are grown from liquid solutions of the elements by epitaxial techniques. This is when a layer, as it solidifies, assumes the crystal structure of the layer beneath it. The process is similar to that used for GaAs and is only slightly more complicated because of the addition of antimony. It is fabricated by first growing a wafer of about 1 square centimeter. The opposite sides are then silvered. Following that, they are diced to yield devices from approximately 0.010 to 0.015 inches.

Electrically, the GaAsSb laser exhibits exceptional dispersion and attenuation characteristics (Fig. 11-1).

EXCITATION OF MOLECULES

A certain amount of energy is required to achieve excitation. The energy is expressed in terms of calories per mole. A mole represents a standard number of molecules.

Electronic excitation requires the highest input of energy, equivalent to that of ultraviolet or visible light. The other forms of excitation require less energy (Fig. 11-2).

Chemical Pumping

Chemical pumping is based on the energy released in the making and breaking of bonds. In Fig. 11-3A, atom A might combine with a molecule consisting of atoms B and C. This would produce an intermediate and transient molecule possessing extra energy. This molecule could separate into two molecular fragments (Fig. 11-3B), Either one might be excited and could be stimulated enough to drop to a lower energy level (Fig. 11-3C), emitting a photon in the process.

Hydrogen and Chlorine Reaction

Reactions between hydrogen and chlorine in an explosion provide pumping for a chemical laser. A trigger of light (Fig.

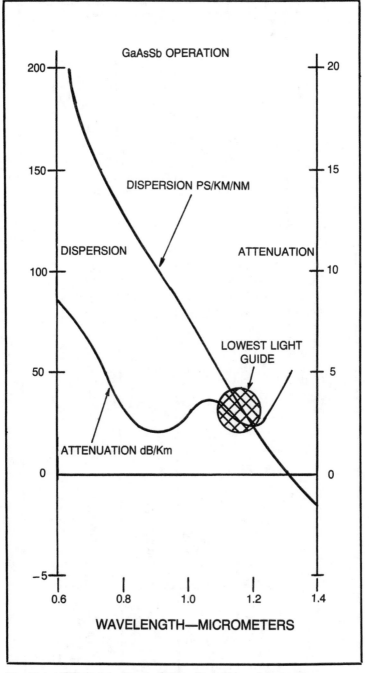

Fig. 11-1. Minimum loss chart exhibiting dispersion and attenuation.

TYPE	EXCITATION	ENERGY LEVEL	MECHANISM
ELEC-TRONIC	VISIBLE OR ULTRAVIOLET LIGHT	100,000 TO 200,000 CALORIES PER MOLE	
VIBRA-TIONAL	INFRARED LIGHT	500 TO 10,000 CALORIES PER MOLE	
ROTA-TIONAL	MICROWAVE LIGHT	0.1 TO 100 CALORIES PER MOLE	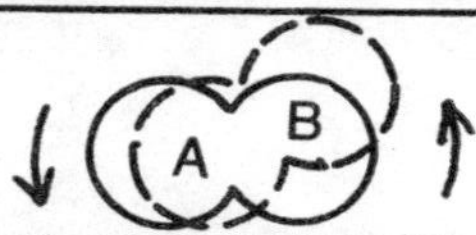
TRANS-LATIONAL	HEAT	ANYTHING ABOVE ZERO	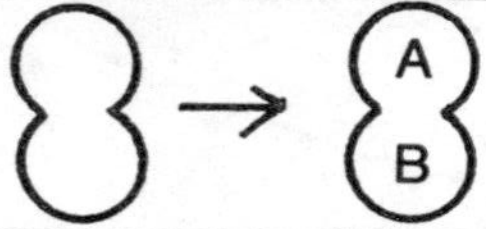

Fig. 11-2. Electronic excitation requires different levels of energy, depending on the type of excitation.

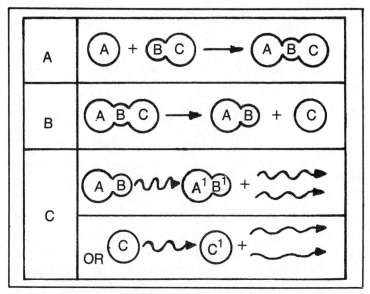

Fig. 11-3. The process of chemical pumping makes and breaks bonds.

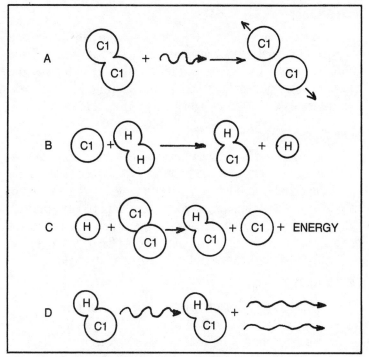

Fig. 11-4. When hydrogen and chlorine react, they produce an excited hydrogen chloride molecule that emits an infrared photon.

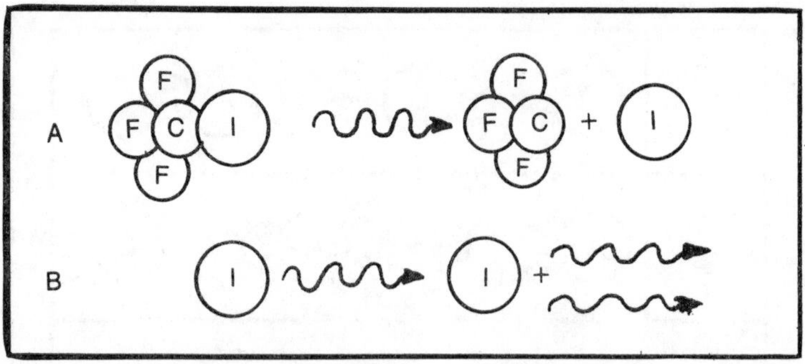

Fig. 11-5. An iodine atom is created when the carbon-iodine bond is ruptured and a photon of light energy is released.

11-4A) separates a chlorine molecule into two chlorine atoms. One of them (Fig. 11-4B) combines with a hydrogen molecule to yield a hydrogen chloride molecule and a free hydrogen atom. This produces an excited hydrogen chlorine molecule (Fig. 11-4C) that then emits a photon of infrared radiation (Fig. 11-4D).

Iodine Laser

The iodine laser derives its pumping from the dissociation by light of a molecule consisting of a carbon atom, three fluorine atoms and an iodine atom (CF3I). In rupture of the carbon-iodine bond, an excited iodine atom (Fig. 11-5) is born. This releases a photon of light energy (Fig. 11-5B).

Liquid Laser Cells

Typical liquid laser cells perform the function of the resonating cavities in solid and gaseous type lasers (Fig. 11-6). All of the liquid lasers have expansion receivers to accommodate the expansion and contraction of the liquid during thermal changes which occur during normal operation. The lasers are optically pumped by means of surrounding flash lamps. These lamps are of either the spiral or the proximity type.

LASER NOTES

Metric Units of Linear Measurement
1 micron = 1 millionth of a meter (10^{-6} meter)
1 millimicron = 1 millionth of a millimeter
1 angstrom = 10^{-8} cm., 10^{-9} meter
1 nanometer = 10^{-9} cm., 10^{-10} meter

Metric—English Units for
the Measurement of Vacuum

one atmosphere = 29.92 inches of Hg. = 760 mm. of Hg.

1 mm. of Hg. = 0.019 psi. = 0.0446 foot of water

1 inch of water = 1.87 mm. of Hg. = 0.036 psi.

1 inch of Hg. = 25.4 mm. of Hg. = 13.6 inches of water

1 mm. of Hg. = 1 torr = 1/760 of an atmosphere

1 inch = 2.54 cm. = 25.4 mm. = 25,400 microns

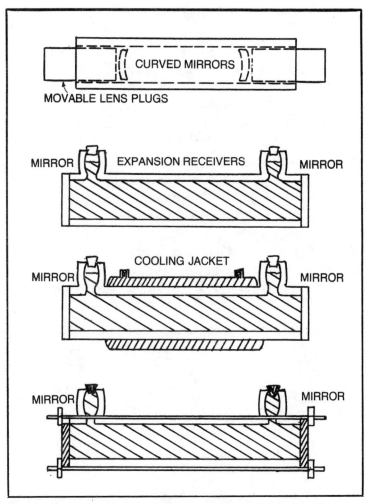

Fig. 11-6. Typical liquid laser cells act as resonating cavities in solid and gaseous type lasers.

Simple Magnifier

$$M = \frac{25 \text{ cm.}}{f}$$

(approximately)

Where f is the focal length of lens.

Diffraction Grating Equation

$$\lambda = \frac{d \text{ sine } \theta}{n}$$

Where λ = wavelength; d is grating constant; θ is diffraction angle; n is the order of image.

Steradian

The Steradian is the subtended solid angle of a spherical portion, where the angle equals the square of the radius (r^2).

$$\pi \text{ steradians} = 1 \text{ steregon}$$

Steradians are units of brilliance or energy factors used to describe power of lasers such as X amount of mw per steradian.

PLANCK'S constant

6.62×10^{-34} joules per second

PLANCK'S formula

$E = hf$

Where E = energy of a photon in joules

f = vibrations per second (Hz.) of radiation

h = Planck's constant: 6.62×10^{-34} joules per second

(6.6×10^{27} erg seconds)

TYPE OF RADIATION	MEAN VALUE OF ϕ (Hz.)	hf IN JOULES
Radio waves	$3 = 10^5$	$2.0 = 10^{-28}$
Heat waves	$3 = 10^{13}$	$2.0 = 10^{-26}$
Visible light	$6 = 10^{14}$	$3.9 = 10^{-19}$
X-rays	$3 = 10^{18}$	$2.0 = 10^{-15}$
Gamma rays	$3 = 10^{19}$	$2.0 = 10^{-14}$

Mass of a Photon

$$m = \frac{hf}{C^2}$$

Where m = the photon mass; h = Planck's constant; f = the frequency in Hz.; C = the speed of light.

Energy of a Photon
$$E = h \times f$$
Where E = the energy in joules; h = Planck's constant; f = the frequency in Hz.

Wavelength of a Photon
$$\lambda = \frac{h}{mc}$$

Where λ = photon wavelength; h = Planck's constant; m = mass of photon; c = the velocity of light (mc is photon momentum).

Wavelength of a Particle Having a Velocity
$$\lambda = \frac{h}{m v}$$

Where λ = the particle wavelength; h is Planck's constant; m = mass of particle; v = velocity (mv is particle momentum).

Einstein's Photoelectric Equation
$\frac{1}{2}mv^2 \text{max} = hf - w$

f = vibrations per second (Hz.)

h = Planck's constant 6.62×10^{-34}

w = the work function of the substance

Speed of Light in Vacuum
300,000,000 meters per second (3.0^8 meters per second)

Speed of light in vacuum = 186,000 miles per second

one kilowatt hour = 3,600,000 joules

1 gm. mass converted to energy = 9×10^{20} ergs = 9×10^{13} joules

Glossary

absolute zero: The lowest possible temperature: -273.16°C. or −459.69°F

absorption spectrum: A continuous spectrum, like white light, interrupted by dark lines or bands that are produced by the absorption of certain wavelengths by a substance through which the light or other radiation passes.

ampere: The M.K.S. unit for measuring electric current (= the flow of one coulomb per second).

amplifier: An electronic device which changes a weak signal into a much stronger one.

amplitude: The maximum displacement, or graphic height, of an oscillating wave during its span. It is measured from *time line* to *peak top*.

Angstrom (A): A unit of length equal to 10^{-10} of a meter, used in measuring the wavelength of light. Atoms have a radius of from one to two angstroms.

atom: The smallest particle of an element that has all of its chemical properties, composed of at least one proton, one neutron and one electron.

atomic number: The number of protons in the nucleus of an atom.

beam divergence in a given plane: The half of divergence of the laser emission at which the intensity of radiation is one half the peak intensity.

beta rays: Streams of fast moving particles (electrons) ejected from radioactive nuclei.

cathode rays: Electrons emitted by a cathode, as in an electron gun.

c.g.s.: The metric system of measurement in which the fundamental units are the centimeter, the gram and the second.

chromatic aberration: The inability of a single lens to refract all the different colors of light to the same focus.

coherent radiation: Radiation in which the phase difference between any two points in the radiation field is constant throughout the duration of the radiation.

color: The property of light which depends on its frequency (Hz). It is visible to the eyes' retina stimulae.

component: One of the several vectors combined algebraically, or geometrically to yield a resultant vector.

concave lens: A lens which diverges parallel light rays to a focal point.

conservation of matter and energy: A law which expresses that the total amount of energy and matter in the universe is constant, which can be equated as $E=mc^2$.

continuous spectrum: A spectrum consisting of a wide range of unseparated wavelengths.

converging lens: A lens that is thicker in the middle than it is at the edge.

convex lens: A lens which converges parallel light rays.

cosmic rays: High energy particles, apparently from beyond our solar system.

coulomb: The quantity of electricity equal to the charge on 6.25×10^{18} of electrons.

constructive interference: The superposition of two waves approximately in phase so that their amplitudes add up to produce a combined wave of larger amplitude than its components.

De Broglie matter waves: All particles of matter have associated wave properties. The wavelength of a particle is related to its momentum and Planck's constant h by the relationship:

$$\text{Wavelength} = \frac{h}{\text{momentum}}$$

decay: The act of an atom or population of atoms falling down from an excited level of energy when the population decays. It is said to revert. Decaying can be thought of as the reverse of stimulation or excitation.

dichroism: A property of certain crystalline substances in which one polarized component of incident light is absorbed, and the other is transmitted.

diffraction grating: An optical surface, either transmitting or reflecting with several thousand equally spaced and parallel grooves ruled in it.

dispersion: The separation of polychromatic light into its component wavelengths.

diverging lens: A lens that is thicker at the edge than it is in the middle.

duty factor (du): The product of the pulse duration and the pulse repetition frequency of a wave composed of pulses that occur at regular intervals.

electromagnetic waves: Transverse waves in space, having an electric component and a magnetic component, each being perpendicular to the other and both being perpendicular to the direction of propagation.

electron: A negatively charged atomic particle having a rest mass of 9.1083 × 10^{-28} g.

electron shell: A region about the nucleus of an atom in which electrons move or orbit.

elementary colors: The six regions of color in the solar spectrum, observed by the dispersion of sunlight, red, orange, yellow, green, blue and violet.

emission: To release or give off, as in releasing or emitting radiant energy. The act of discharging excess energy.

energy level: A region about the nucleus of an atom, where electrons are orbiting, also representing the electrons particular energy status.

erg: The C.G.S. system unit of work, a force of one dyne acting through a distance of one centimeter.

excitation: The process of boosting one or more electrons in an atom or molecule from a lower to a higher energy level. An atom in such an excited state will usually decay rapidly to a lower state or level accompanied by the emission of radiation. The frequency and energy of emitted radiation are related by: $E = hf$.

F number: The ratio of the focal length of a lens to the effective aperture.

focal length: The distance between the principal focus of a lens and its optical center or vertex.

focus: A point at which light rays meet or from which rays of light diverge.

focus, principal: A point to which rays parallel to the principal axis converge, or from which they diverge, after reflection or refraction.

Fraunhofer lines: Absorption lines in the solar spectrum.

frequency: Number of vibrations or cycles per second (c.p.s. or Hertz (Hz).

frequency, cut-off: A characteristic threshold frequency of incident light, below which, for a given material, the photoelectric emission of electrons ceases.

gamma ray: A high energy wave emitted from the nucleus of a radioactive atom.

ground level: That energy level of the atomic population which occurs naturally or in absence of additive stimulation. The relative level before population inversion is affected, from which the level of excitation after inversion may be measured or compared from is the base line or reference point in considering the degree of excitation.

index of refraction: The ratio of the speed of light in a vacuum to its speed in any other given substance.

infrared-emitting diode: A semiconductor device in which radiation recombination of injected minority carriers produce infrared radiatn flux when current flows as a result of applied voltage.

injection laser: A solid-state semiconductor device consisting of at least one p-n junction capable of emitting coherent or stimulated radiation under specified conditions. The device will incorporate a resonant optical cavity.

interference: The superposing of one wave on another, in either a constructive or destructive manner. The mutual effect of two beams of light, resulting in a loss of energy in certain areas and reinforcement of energy in others.

joule: The M.K.S. unit of work, a force of one newton acting through a distance of one meter (*nt-m*).

kilowatt hour: A unit of electric energy equal to: 3.6×10^6 watts per seccond.

laser: A device which emits coherent monochromatic light by the process of *light amplification by the stimulated emission of radiation.* An optic maser. —A device which emits coherent, amplified radiations on the sub-visible range. Being in the microwave category, accomplished by *microwave amplification by the stimulated emission of radiation,* the maser was invented first and the laser was a ramification of it.

lasing condition (or state): The condition of an injection laser corresponding to the emission of predominantly coherent or stimulated radiation.

mass: The measure of the quantity of matter.

matter: Anything that occupies space and has weight.

meter: The main, or basic unit of length in the metric system (39.37 inches).

molecule: The smallest particle that an element or compound can exist as and be chemically independent.

mole: That quantity of a substance whose mass in grams is numerically equal to the mass of one of its molecules in atomic mass units.

nanometer: Unit of measurement sometimes used in light technology. It is equal to 10^{-9} M. or .00000003937 inches (nm.).

neutron: A neutral atomic particle having a mass of 1.675×10^{-24} grams.

nucleus: The positively charged dense central part of an atom.

optical center: The point in a lens through which the secondary axis passes.

optical density: A property of a transparent material which is a measure of the speed of light through it.

paramagnetism: The property of a substance bv which it is attracted by a strong magnet.

phase: The position and motion of a particle of a wave.

photoelectric effect: The emission of electrons by a substance when illuminated by electromagnetic radiation.

photoelectric effect, inverse: The emission of photons of radiation due to the bombardment of a material with high speed electrons.

photoelectric emission, 1st. law of: The rate of emission of photoelectrons is directly proportional to the intensity of the incident light.

photoelectric emission, 2nd. law of: The kinetic energy of photoelectrons is independent of the intensity of the incident light.

photoelectric emission, 3rd. law of: Within the region of effective frequencies, the maximum kinetic energy of photoelectrons is directly proportional to the frequency of the incident light.

photoelectrons: Electrons emitted from a light sensitive material when it is properly illuminated.

photon: A quantum of light energy whereby using Planck's constant h relates the energy and frequency of light quanta: $h = 6.6 \times 10^{-34}$ Joules per second. Photons are also thought of as being in the form of packets of energy.

piezoelectric effect: The property of certain natural and synthetic crystals to develop a potential difference (voltage) between opposite surfaces when subjected to mechanical stress or vibrations.

Planck's constant: A universally proportional constant, relating photon energy to the frequency of radiation ($h = 6.6 \times 10^{-34}$ joules per second).

plasma: Hot gases composed of electrically charged particles. Most of the matter in the universe is in the plasma state.

population: The total, or sum of the atoms in a mass.

population inversion: The act or process of inverting the population or atoms in a mass from one state or condition to that of another. To raise the normally occurring small percentage of the atomic population which exists in an excited state, to a greater majority or percentage of this population to a higher level of excitation, thus inverting the percentage or ratio of the population from a lower to a higher level.

population reversion: The condition or event which exhibits the atomic decay of the population to its former energy level, before the population was first inverted. A return to its original level uninverted state, accompanied by the release or emission of the previously applied or stimulated energy. It is emitted in the form of a packet of energy, also known as a photon.

prelasing condition (or state): The condition of an injection laser corresponding to the emission of predominantly incoherent or spontaneous radiation.

primary colors: Red, blue and green. From these, all other secondary colors may be achieved.

principal axis: A line drawn through the center of curvature and the optical center of a lens.

proton: A positively charged atomic particle having a mass of 1.67×10^{-4} grams.

pulse: A singularly non-recurring disturbance, such as a pulse laser.

quantum efficiency (QE): The quantum efficiency of a source of radiation flux is the ratio of the number of quanta of radiant energy (photons) emitted per second to the number of electrons flowing per second, e.g., photons/electron.

quantum theory: The theory that the transfer of energy between light radiations and matter occurs in discrete units or packets, the magnitude depending on the frequency of the radiation.

radiant efficiency of a source of radiant flux (η): The radiant efficiency of a source of radiant flux is the ratio of the total radiant flux to the forward power dissipation.

radiant flux (radiant power) (Φ): Radiant flux is the time rate of flow of radiant energy. It is expressed preferably in watts, or ergs/second.

Radiant Intensity (I): The radiant intensity of a source is the radiant flux proceeding from the source per unit solid angle in the direction considered, e.g., watts/steradian.

radiation: The form of energy in transit that occurs when an atom decays, or reverts to ground level after radiating the necessary energy for it to do so. Radiation can, and quite often does, manifest itself as visible radiation or light in the form of photons, or discrete packets of energy.

radiation pattern: The representation of the intensity of emission as a function of direction, in a given plane. The axes are to be specified with respect to the junction plane and the cavity face.

rectilinear propagation: Energy waves travelling in a straight line.

refraction: The bending of a wave disturbance as it passes obliquely from one medium into another of different density.

resultant: A vector representing the algebraic or geometric sum of several components.

resonance: (1) The inducting of vibrations or oscillations of a natural rate in matter by a vibrating or oscillating source, having the same or a multiple related frequency. (2) The condition whereby energy waves may be amplified by the continued production of waves having a duplicated length as the cavity length where such waves are produced in. Such resonant cavities are utilized in the maser and laser.

rise time (t_r): The time taken for the radiation flux to increase from 10% to 90% of its peak value when the laser is subjected to a step function current pulse of specified amplitude.

scattering: When light falls on a medium, electrons in the medium are set into oscillation by the time-varying electric vector of the incident light. The electrons in turn, emit light in every direction, scattering the beam (incident) from its original straight line path.

secondary axis: Any line, other than the principal axis drawn through the center of curvature of the optical center of a lens.

secondary emission: Emission of electrons as a result of the bombardment of an electrode by high velocity electrons.

spectral bandwidth ($\Delta\lambda$): The spectral bandwidth for single peak devices is the difference between the wavelengths at which the radiant intensity is 50% (unless otherwise stated) of the maximum value.

spectral radiant flux ($\Phi\lambda$): Spectral radiant flux is the radiant flux per unit wavelength interval at wavelength λ, e.g., watts/nanometer.

stimulation: To cause something or some atom to increase or decrease its value in some respect. To trigger a reaction in an atom.

superposition: The process of combining the displacements of two or more wave motions, algebraically to produce a resultant wave motion.

thermionic emission: The liberation of electrons from the surface of a heated body, usually occurring at a point of visible incandescence.

threshold current (Ith): The minimum forward current for which the laser is in a lasing state at a specified temperature.

threshold frequency: The minimum frequency of incident light that will eject a photon from a given metal or substance.

transverse wave: A wave in which the particles of the medium vibrate at right angles to the path along which the wave travels through the medium.

Vander Waals forces: Attractive forces arising from the effect of the varying electric field of atoms of one molecule on the electric field of atoms of another molecule.

vector quantity: A quantity which requires both a magnitude and a direction for its complete description.

watt: The M.K.S. system unit of power (one joule per second, 1V. ×1A).

wavelength: The distance from one particle or peak to the next following particle or peak measured along a parallel line with the time line.

wavelength of peak radiant intensity: The wavelength at which the spectral distribution of radiant intensity is a maximum.

x-rays: Electromagnetic radiations of very short wavelengths and high frequency, enabling great penetrating power into and through substahces.

Appendix A
The Periodic Table

The modern Periodic Table, which differs only slightly from the table as compiled by Dimitri Mendeleev in 1869, is a tabular description of all of the naturally occurring elements (92) and the additional 14 synthesized man-made elements. These are a total of 106 occupations in the Periodic System.

Reference will be made to the Periodic Table and to Figs. A-1 and A-2. For purposes of explanation, Iron (Fe) has been selected as an example of what the Periodic Table tells about each and every element of the system.

In the upper left hand corner (Fig. A-1) will be a number ranging from 1 to 103. This number is the *atomic number* of the element. It tells us two things about the element: the number of protons in the nucleus and the number of electrons that the atom will normally have in a neutral state. There will be present one electron in the outer orbital envelope for each proton present in the nucleus.

In the center of the box there will be found the *chemical symbol* or abbreviation for the element. In this case iron is chemically symbolized as *Fe*.

Below the chemical symbol will be the *atomic weight* of the atom. This weight represents the weight of the element. It considers its mass which consists of its nucleon members, or number of protons and neutrons. Thus, we may very easily

Table A-1. Periodic Table.

PERIOD / COLUMN	I	II	21	22	23	24	25	26	27	28	29	30	III	IV	V	VI	VII	0	ORBITALS BEING FILLED
n = 1	1 H 1.00																		2 He 4.00	1s
n = 2	3 Li 6.94	4 Be 9.01												5 B 10.8	6 C 12.01	7 N 14.01	8 O 16.00	9 F 19.0	10 Ne 20.2	2s2p
n = 3	11 Na 23.0	12 Mg 24.3												13 Al 27.0	14 Si 28.1	15 P 31.0	16 S 32.1	17 Cl 35.5	18 Ar 39.9	3s3p
n = 4	19 K 39.1	20 Ca 40.1	21 Sc 45.0	22 Ti 47.9	23 V 50.9	24 Cr 52.0	25 Mn 54.9	26 Fe 55.8	27 Co 58.9	28 Ni 58.7	29 Cu 63.5	30 Zn 65.4	31 Ga 69.7	32 Ge 72.6	33 As 74.9	34 Se 79.0	35 Br 79.9	36 Kr 83.8	4s3d4p
n = 5	37 Rb 85.5	38 Sr 87.6	39 Y 88.9	40 Zr 91.2	41 Nb 92.9	42 Mo 95.9	43 Tc (99)	44 Ru 101.1	45 Rh 102.9	46 Pd 106.4	47 Ag 107.9	48 Cd 112.4	49 In 114.8	50 Sn 118.7	51 Sb 121.8	52 Te 127.6	53 I 126.9	54 Xe 131.3	5s4d5p
n = 6	55 Cs 132.9	56 Ba 137.3	57-71 See Below	72 Hf 178.5	73 Ta 180.9	74 W 183.9	75 Re 186.2	76 Os 190.2	77 Ir 192.2	78 Pt 195.1	79 Au 197.0	80 Hg 200.6	81 Tl 204.4	82 Pb 207.2	83 Bi 209.0	84 Po (209)	85 At (210)	86 Rn (222)	6s4f5d6p
n = 7	87 Fr (223)	88 Ra (226)	89- See Below	Below																7s5f6d7p

TRANSITION ELEMENTS

	57	58	59	60	61	62	63	64	65	66	67	68	69	70	71	ORBITALS BEING FILLED
n = 6	La 138.9	Ce 140.1	Pr 140.9	Nd 144.2	Pm (147)	Sm 150.4	Eu 152.0	Gd 157.3	Tb 158.9	Dy 162.5	Ho 164.9	Er 167.3	Tm 168.9	Yb 173.0	Lu 175.0	4f
	89	90	91	92	93	94	95	96	97	98	99	100	101	102	103	
n = 7	Ac (227)	Th (232)	Pa (231)	U 238.0	Np (237)	Pu (242)	Am (243)	Cm (247)	Bk (249)	Cf (251)	Es (254)	Fm (253)	Md (256)	No (253)	Lw (257)	5f

PARENTHETICAL VALUES ARE MASS NUMBERS OF THE ISOTOPES WITH LONGEST HALF LIVES

182

determine the number of neutrons of the element by rounding off the atomic weight to the nearest whole number and subtracting from this atomic weight, the number of protons as given in the atomic number. Our answer will be the number of neutrons in the nucleus.

Referring to the Periodic Table (Table A-1), we observe on the far left-hand side, the notation *PERIOD COLUMN*. This tells us the number of electron orbits the atom has in its outer envelope. It will be seen that iron (Fe) has four such orbits. On the box side to the right of the box is provided the number of electrons in each individual orbit. This enables us to construct a geometric drawing of each element from all of the notations given.

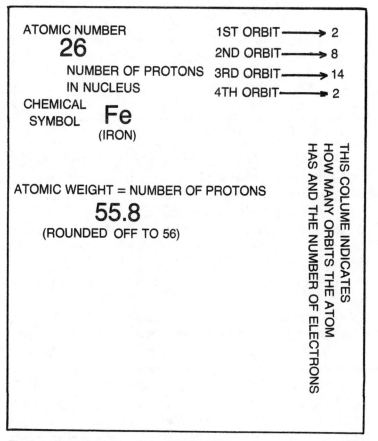

Fig. A-1. Description of iron in the Periodic Table.

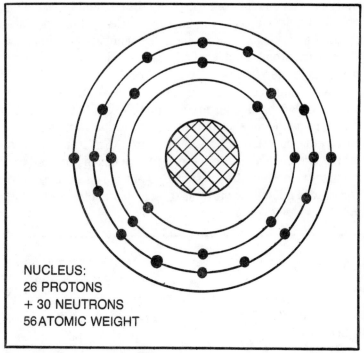

NUCLEUS:
26 PROTONS
+ 30 NEUTRONS
56 ATOMIC WEIGHT

Fig. A-2. The atomic geometry of iron.

The number of electrons in the outermost orbit also tells us the valence of the element. The combining ability of each atom is governed by the number of electrons, or hooks in its outer orbit. Thus with iron having an outer number of two electrons, it then has a valence of two. Looking toward the right-hand side of the Table will be seen a stair-step like division of the elements. This has been added to separate the elements from the metals and nonmetals. All those 84 elements to the left of the division are metals and those to the right are nonmetals. To the extreme right, under the heading 0, are the inert elements which by virtue of their completed outer orbits, will not combine with any elements whatsoever. These elements are referred to as *chemically inert*.

The original Table, as compiled Mendeleev had the elements arranged in successive order by their atomic weights. The modern revision arranges the progression according to the element's proton count which is a more practical arrangement for scientific purposes.

ATOMIC WEIGHTS OF THE ELEMENTS

Element	Symbol	Atomic Number	Atomic Weight	Element	Symbol	Atomic Number	Atomic Weight
Actinium	Ac	89	(227)	Mercury	Hg	80	200.59
Aluminum	Al	13	26.98	Molybdenum	Mo	42	95.94
Americium	Am	95	(243)	Neodymium	Nd	60	144.24
Antimony	Sb	51	121.75	Neon	Ne	10	20.183
Argon	Ar	18	39.948	Neptunium	Np	93	(237)
Arsenic	As	33	74.92	Nickel	Ni	28	58.71
Astatine	At	85	(210)	Niobium	Nb	41	92.91
Barium	Ba	56	137.34	Nitrogen	N	7	14.007
Berkelium	Bk	97	(249)	Nobelium	No	102	(253)
Beryllium	Be	4	9.012	Osmium	Os	76	190.2
Bismuth	Bi	83	208.98	Oxygen	O	8	15.9994
Boron	B	5	10.81	Palladium	Pd	46	106.4
Bromine	Br	35	79.909	Phosphorus	P	15	30.974
Cadmium	Cd	48	112.40	Platinum	Pt	78	195.09
Calcium	Ca	20	40.08	Plutonium	Pu	94	(242)
Californium	Cf	98	(251)	Polonium	Po	84	(210)
Carbon	C	6	12.011	Potassium	K	19	39.102
Cerium	Ce	58	140.12	Praseodymium	Pr	59	140.91
Cesium	Cs	55	132.91	Promethium	Pm	61	(147)
Chlorine	Cl	17	35.453	Protactinium	Pa	91	(231)
Chromium	Cr	24	52.00	Radium	Ra	88	(226)
Cobalt	Co	27	58.93	Radon	Rn	86	(222)
Copper	Cu	29	63.54	Rhenium	Re	75	186.23
Curium	Cm	96	(247)	Rhodium	Rh	45	102.91
Dysprosium	Dy	66	162.50	Rubidium	Rb	37	85.47
Einsteinium	Es	99	(254)	Ruthenium	Ru	44	101.1
Erbium	Er	68	167.26	Samarium	Sm	62	150.35
Europium	Eu	63	151.96	Scandium	Sc	21	44.96
Fermium	Fm	100	(253)	Selenium	Se	34	78.96
Fluorine	F	9	19.00	Silicon	Si	14	28.09
Francium	Fr	87	(223)	Silver	Ag	47	107.870
Gadolinium	Gd	64	157.25	Sodium	Na	11	22.9898
Gallium	Ga	31	69.72	Strontium	Sr	38	87.62
Germanium	Ge	32	72.59	Sulfur	S	16	32.064
Gold	Au	79	196.97	Tantalum	Ta	73	180.95
Hafnium	Hf	72	178.49	Technetium	Tc	43	(99)
Helium	He	2	4.003	Tellurium	Te	52	127.60
Holmium	Ho	67	164.93	Terbium	Tb	65	158.92
Hydrogen	H	1	1.0080	Thallium	Tl	81	204.37
Indium	In	49	114.82	Thorium	Th	90	232.04
Iodine	I	53	126.90	Thulium	Tm	69	168.93
Iridium	Ir	77	192.2	Tin	Sn	50	118.69
Iron	Fe	26	55.85	Titanium	Ti	22	47.90
Krypton	Kr	36	83.80	Tungsten	W	74	183.85
Lanthanum	La	57	138.91	Uranium	U	92	238.03
Lawrencium	Lw	103	(257)	Vanadium	V	23	50.94
Lead	Pb	82	207.19	Xenon	Xe	54	131.30
Lithium	Li	3	6.939	Ytterbium	Yb	70	173.04
Lutetium	Lu	71	174.97	Yttrium	Y	39	88.91
Magnesium	Mg	12	24.312	Zinc	Zn	30	65.37
Manganese	Mn	25	54.94	Zirconium	Zr	40	91.22
Mendelevium	Md	101	(256)				

Based on mass of C^{12} at 12.000. Values in parentheses represent the most stable known isotopes for elements which do not occur naturally.

Appendix C
Laser Parts Suppliers

Beam Splitters
C.V.I. Laser Corp.
P.O. Box 11308
Albuquerque, NM 87112
Phone: 505-296-9541

Crystal Brewster Windows
Harshaw Chemical Co.
Broad St.
Gloucester, NJ 08030

Crystals
Adolf Meller Co.
P.O. Box 6001
Providence, RI 02940

Metrologic Instruments Inc.
143 Harding Avenue
Bellmar, NJ 08030
Phone: 609-933-0100

Coatings
Herron Optical, Div. Bausch & Lomb
2035 East 223rd Street
Long Beach, CA 90810
Phone: 213-830-5404

Components
Coherent Radiation Co.
3210 Porter Drive
Palo Alto, CA 94304

Power Tech. Inc.
P.O. Box 4403
Little Rock, AR 72214
Phone: 501-568-1195

Diode Pulsers
Power Tech. Inc.
P.O. Box 4403
Little Rock, AR 72214
Phone: 501-568-1995

Dyes
ExCiton Company
5760 Burkhardt Rd.
Dayton, OH 45431
Phone: 513-252-2989

Electronic Components
Havilton Electro-Sales
340 Middlefield Rd.
Mt. View, CA 94041

Mazda Electronics
1287 Lawrence Blvd.
Sunnyvale, CA 94086

Metrologic Instruments Inc.
143 Harding Avenue
Bellmar, NJ 08030
Phone: 609-933-0100

Electronic Parts
Newark Electronics
500 N. Pulaski Rd.
Chicago, IL 60624

Radio Shack
2617 W. 7 th. Street
Ft. Worth, TX 79901

Electro-Optical Components

*R.C.A. Radio Corporation
of America*
Electro-Optics Div.
Lancaster, PA 17604
Phone: 717-397-7661
(RC.A. L.E.D. Diode,
#FLV-104 and S-6200
series diodes.)

Experimental Parts
Radio Shack
2617 W. 7th Street
Ft. Worth, TX 79901

Fabry-Perot Holders
C.V.I. Laser Corp.
P.O. Box 11308
Albuquerque, NM 87112
Phone: 505-296-9541

Fused Quartz Lens & Prisms
*Thermal American Fused
Quartz Co.*
Route 202 & Change Bridge Rd.
Montville, NJ 07045
Phone: 201-334-7770

Gases
Matheson Gas Co.
Box 85
East Rutherford (Box 85),
NJ 07073
(Gases for science, CO_2, He,
Neon, Nitrogen, Argon,
Hydrogen, etc.)

Gas Lasers
*Hughes Aircraft Company,
Industrial Products Div.*
6155 El Cavino Real
Carlsbad, CA 92008
Phone: 714-438-9191

General Laser Parts
Cenco Scientific
2600 S. Kostner Ave.
Chicago, IL 60623
Phone: 312-277-8300

Ealin Corp.
22 Pleasant St.
South Natich, MA 01760
Phone: 617-655-7000

Edmond Scientific Co.
Dept. B-09, Edscorp Bldg.
Barrington, NJ 08007

Esco Products
Oak Ridge Rd.
Oak Ridge, NJ 07438

Fisher Scientific
711 Forbes Ave.
Pittsburgh, PA 15219

Information Unlimited
Box 716
Amherst, NH 03031
Phone: 603-373-4730

Lasermetrics
Teaneck, NJ 07666
Phone: 201-837-9090

Pioneer Industries
10-A Haughey St.
Nashua, NH 03060
Phone: 603-882-7215

Sargent Welch
7300 N. Linden Ave.
Skokie, IL 60076
Phone: 312-677-0600

Solaser
Box 1005
Claremont, CA 91711

Spectra-Physics
1250 West Middlefield Road
Mountain View, CA 94040

V.W.R. Scientific
P.O. Box 1050
Rochester, NY 14603
Phone: 212-294-3000

Glass Products
Arthur H. Thomas Co.
3rd and Vine Street
Philadelphia, PA 19105

Plasma Scientific Co.
P.O. Box 801
Cucamonga, CA 91730

Information & Data
United Electronics Institute
3947 Park Dr.
Louisville, KY 40216

Lasers (Assembled)
*Hughes Aircraft Company,
Industrial Products Div.*
6155 El Camino Real
Carlsbad, CA 92008
Phone: 714-438-9191

Metrologic Instruments Inc.
143 Harding Avenue
Bellmar, NJ 08030
Phone: 609-933-0100

Lenses
Lambda Optics
Berkley, NJ 07922
Phone: 201-464-5060

Mechanical Positioning Devices
Burleigh Instruments Inc.
100 Despatch Drive, Box 270
East Rochester, NY 14445
Phone: 716-586-7930

Mirror Kits
C V.I. Laser Corp.
P.O. Box 11308
Albuquerque, NM 87112
Phone: 505-296-9541

*Herron Optical, Div.
Bausch & Lomb*
2035 East 223rd. Street
Long Beach, CA 90810
Phone: 213-830-5404

P. T. R. Optics
145 Newton Street
Waltham, MA 02154
Phone: 617-891-6000

Optical & Lens Supplies

Herron Optical, Div.
Bausch & Lomb
2035 East 223rd. Street
Long Beach, CA 90810
Phone: 213-830-5404

Lambda Optics
Berkley, NJ 07922
Phone: 201-464-5060

Power Supplies

Power Tech. Inc.
P.O. Box 4403
Little Rock, AR 72214
Phone: 501-568-1995

Raytheon Comapny
28 Seyon Street
Waltham, MA 02154

Pyrex Parts

Dowel Corning Co.
(Nationwide—consult your
telephone directory).

Q-Switches

Cleveland Crystals Inc.
19306 Redwood Ave.
Cleveland, OH 44110
Phone: 216-486-6100

Ruby Rods

Adolf Meller Co.
P.O. Box 6001
Providence, RI

Simulators

Power Tech. Inc.
P.O. Box 4403
Little Rock, AR 72214
Phone: 501-568-1995

Special Components

Adolf Meller Co.
P.O. Box 6001
Providence, RI 02940

Systems

Power Tech. Inc.
P.O. Box 4403
Little Rock, AR 72214
Phone: 501-568-1995

Tubes

Coherent Radiation Co.
3210 Porter Drive
Palo Alto, CA 94304

Raytheon Company
28 Seyon Street
Waltham, MA 02154

Wave Plates

C.V.I. Laser Corp.
P.O. Box 11308
Albuquerque, NM 87112
Phone: 505-296-954

Appendix D
Tabulation of Laser Data

MATERIAL	TEMP. deg. K	PUMP REGION	WAVELENGTH	THRESHOLD
(SYMBOL)		(A)	(microns)	(joules)
$CaWO_4:Nd^{3+}$	77	5700-6000	A 1.0650	1.50
			B 1.0633	14.00
			C 1.0660	6.00
			D 1.0576	80.00
			E 1.0641	7.00
	295		A 1.0652	3.00
			D 1.0582	2.00
$SrWO_4;Nd^{3+}$	77	5700-6000	A 1.0574	4.70
			B 1.0627	5.10
			C 1.0607	7.60
	295		B 1.0630	180.00
$SrMoO_4:Nd^{3+}$	77	5700-6000	A 1.0640	17.00
			B 1.0652	70.00
			C 1.0590	150.00
			D 1.0627	170.00
			E 1.0611	500.00
	295		A 1.0643	125.00
			F 1.0576	45.00
$CaMoO_4;Nd^{3+}$	77	5700-5900	1.0670	100.00
	295		1.0673	360.00
$PbMoO_4Nd^{3+}$	295	5700-5900	1.0586	60.00
$CaF_2:Nd^{3+}$	77	7000-8000	1.0457	60.00
		5600-5800		
$SrF_2:Nd^{3+}$	77	7200-7500	1.0437	150.00
	295	7800-8100	1.0370	480.00
$BaF_2:Nd^{3+}$	77	5700-6000	1.0600	1600.00
$LaF_3:Nd^{3+}$	77	5000-6000	A 1.0631	93.00
			B 1.0399	75.00
	295		A 1.0633	150.00
$CaWO_4:Ho^{3+}$	77	4400-4600	2.0460	80.00
			2.0590	250.00
$CaF_2:Ho^{3+}$	77	4000-6600	2.0920	260.00
$CaWO_4:Tm^{3+}$	77	4600-4800	1.9110	60.00
		17000-18000	1.9160	73.00
$SrF_2:Tm^{3+}$	77		1.9720	1600.00
$CaF_2:Tm^{2}+$	20	2800-3400	1.1153	450.00
		3900-4600		
		5300-6300		
	77		1.1153	800.00

Appendix E
Laser Technical Data

ACTIVE MATERIAL AND VALANCE	OUTPUT WAVELENGTH	HOST MATERIAL	OPERATING MODE	OPERATING TEMPERATURE C.	LASER TYPE
EUROPIUM (3+)	0.61	ytrium oxide plastic chelate in alcohol	PULSED	20	SOLID OR IONIC LASERS
CHROMIUM (3+)	0.70	aluminum oxide	CONTINUOUS	20	SOLID OR IONIC LASERS
SAMARIUM (2+)	0.71	flourides of calcium stronium	CONTINUOUS	20	SOLID OR IONIC LASERS
YTTERBIUM (3+)	1.02	glass	PULSED	-196	SOLID OR IONIC LASERS
PRASEODYNIUM (3+)	1.05	calcium-tungstate	PULSED	-196	SOLID OR IONIC LASERS
NEODYMIUM (3+)	1.06	various flourides, molybdates and glass	PULSED / CONTINUOUS	-196 / 20	SOLID OR IONIC LASERS
THULIUM (2+)	1.12	calcium flouride	PULSED	-253	SOLID OR IONIC LASERS
ERBIUM (3+)	1.61	calcium-tungstate	PULSED	-196	SOLID OR IONIC LASERS
THULIUM (3+)	1.91	cal-tungstate stron-flouride	PULSED	-196	SOLID OR IONIC LASERS
HOLMIUM (3+)	2.05	cal-flouride cal-tungstate and glass	PULSED	-196	SOLID OR IONIC LASERS
DYSPROSIUM (2+)	2.36	cal-flouride			SOLID OR IONIC LASERS
URANIUM (3+)	2.4–2.6	poly-flourides			SOLID OR IONIC LASERS
HELIUM	160 wavelengths, between 5,940 angstrom units (0.594 micron) and 35 microns		CONTINUOUS		GAS DISCHARGE LASERS
NEON			CONTINUOUS		GAS DISCHARGE LASERS
KRYPTON			CONTINUOUS		GAS DISCHARGE LASERS
XENON			CONTINUOUS		GAS DISCHARGE LASERS
CARBON MONOXIDE			CONTINUOUS		GAS DISCHARGE LASERS
OXYGEN			CONTINUOUS		GAS DISCHARGE LASERS
OTHER GASES			CONTINUOUS		GAS DISCHARGE LASERS
GALLIUM— ARSENIDE PHOSPHIDE	0.65–0.84		PULSED	-175	SEMI-CONDUCTOR INJECTION LASERS
GALLIUM ARSENIDE	0.84		PULSED	20	SEMI-CONDUCTOR INJECTION LASERS
			CONTINUOUS	-196	SEMI-CONDUCTOR INJECTION LASERS
INDIUM PHOSPHIDE	0.91		PULSED	-153	SEMI-CONDUCTOR INJECTION LASERS
			CONTINUOUS	-253	SEMI-CONDUCTOR INJECTION LASERS
INDIUM ARSENIDE	3.1		PULSED	-196	SEMI-CONDUCTOR INJECTION LASERS
			CONTINUOUS	-296	SEMI-CONDUCTOR INJECTION LASERS

Appendix F
Index of Refraction

The *index of refraction* (Table F-1) represents the speed of light through a substance or material and is compared to the speed of light through a vacuum.

Snell's Law of Refraction relates that: If n represents the index of refraction, i the angle of incidence and r the angle of refraction, then:

$$n = \frac{\sin\text{-}i}{\sin\text{-}r}$$

Table F-1. Index of Refraction for Materials.

VACUUM	1.000
AIR at 0 degrees C	1.00029
AIR at 30 degrees C	1.00026
WATER at 50 degrees C	1.330
ICE at 0 degrees C	1.310
CARBON TET	1.460
DIAMOND	2.470
GLASS, (CROWN)	1.510
GLASS, (FLINT)	1.710
GLYCERINE	1.470
ALCOHOL (ETHYL)	1.360
BENZENE	1.500
CARBON DIOXIDE	1.00045
QUARTZ, (FUSED)	1.460

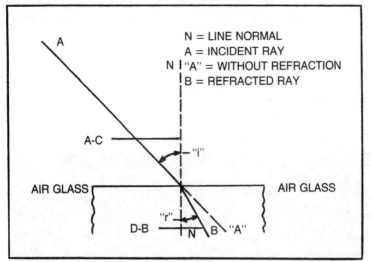

Fig. F-1. Angle of refraction.

Where,
"sin" equals the trigonometric sine of the angle (Fig. F-1).

Angle of Refraction: $\dfrac{\sin i}{\text{A-C}}$ equals $\dfrac{\sin r}{\text{D-B}}$

Example of

Index of Refraction: $\dfrac{\text{speed of light (vacuum)}}{\text{speed of light (glass)}} =$

$\dfrac{186{,}000}{124{,}000} = 1.50$

Appendix G
RCA IR Emitters
and Injection Lasers

This appendix provides data about RCA's standard line of infrared-emitting diodes, single-diode injection lasers, stacked diode lasers, laser arrays, and optically-coupled isolators. More complete data is contained in individual bulletins for the different devices.

Special attention is called to the gallium aluminum arsenide IR emitters and CW-operated injection lasers. These devices are designed primarily for use in optical communication systems. Ratings and characteristics for the different units that are currently available are emphasized by shaded areas in the tabulated data sections.

The type C30133 exemplifies RCA's efforts in this application area. This device, which is the prototype of a planned family, has an integral single-fiber optical cable to obviate user source-to-fiber coupling problems. Variants of this device having different optical waveguide specifications and types having detector coupling units attached to the light pipe are under development.

A cutaway of the RCA C30133 is shown on page 195.

General Information

RCA offers two distinct types of semiconductor photon-emitting devices: infrared-emitting diodes and injection lasers.

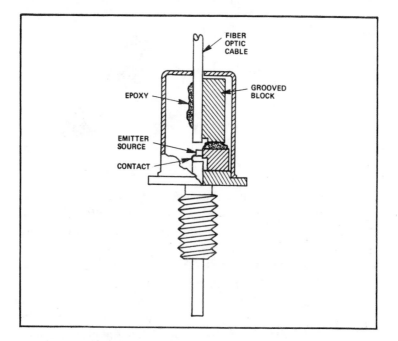

FIBER OPTIC CABLE

EPOXY

GROOVED BLOCK

EMITTER SOURCE

CONTACT

Infrared-Emitting Diodes

Infrared-emitting diodes are p-n diodes in which a fraction of the injected minority carriers recombine by means of radiative transitions. When the junction is forward biased, electrons from the n region are injected into the p region where they recombine with excess holes. In the radiative process, energy given up in recombination is in the form of photon emission. The generated photons travel through the lattice until they are either re-absorbed by the crystal or escape from the surface as radiant flux.

The general structure of a semiconductor IR-emitting diode is shown in Fig G-1.

This type of diode may be operated in either the DC (CW) or pulsed mode. Figures G-2 and G-3 compare radiant flux output from a typical IR emitter for both modes of operation.

The primary pellet material used in RCA IR-emitting diodes is gallium arsenide (GaAs); its wavelength of peak emission is 940 nanometers. When aluminum is added to gallium arsenide (GaAlAs), the output wavelength can be controlled to peak within the spectral range of 800 to 900

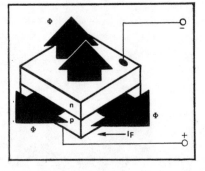

Fig. G-1. General structure of a semiconductor IR-emitting diode.

nanometers. When indium is added (InGaAs), the peak wavelength may be shifted to 1060 nanometers. Typical spectral emission curves for these three pellet materials are shown in Fig. G-4.

Typical structures of RCA IR-emitters are shown in Fig. G-5 and G-6. Both structures have integral collimating systems which provide relatively narrow output beam patterns. For applications requiring a broad output beam pattern, the glass lens shown in Fig. G-6 is replaced by a flat glass window. Figure G-7 shows typical beam patterns for both packages having internal collimating systems and packages having flat glass windows.

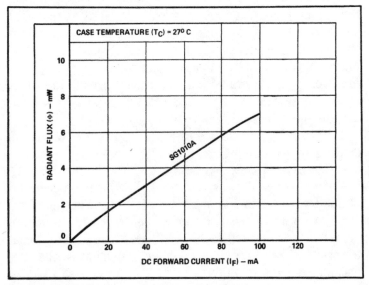

Fig. G-2. Typical radiant flux vs DC forward current.

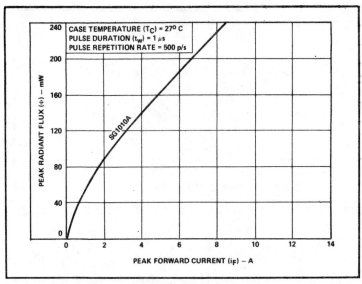

Fig. G-3. Typical peak radiant flux vs peak forward current.

The structures illustrated in Figs. G-5 and G-6 are combination edge and surface emitters and accordingly have relatively large source sizes. Some applications, such as optical communications, require extremely small sources to insure efficient coupling to light pipes or single-fiber optical

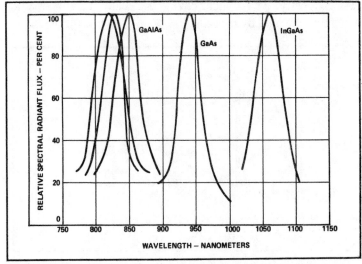

Fig. G-4. Typical spectral emission characteristics of RCA IR-emitters.

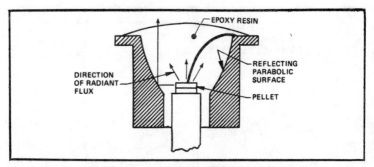

Fig. G-5. Schematic arrangement of a typical RCA IR-emitting diode mounted in an OP-10 package (scale exaggerated).

waveguides. Small course sizes are achieved by sawing the pellet into relatively narrow widths and using only the emission from a single face. The resulting structure is an edge emitter. This type of structure is incorporated in RCA's GaAlAs infrared emitters. The source size of these devices is typically 1 × 6 mils.

Typical life test data for RCA IR-emitting diodes are shown in Fig. G-8. Both CW and pulse operating modes are shown.

Injection Lasers. Laser diodes made from direct bandgap materials differ from conventional infrared-emitting diodes in that they require an optical cavity and a high injection carrier density. The optical cavity is usually a Fabry-Perot cavity type formed by cleaving the opposite ends of the diode

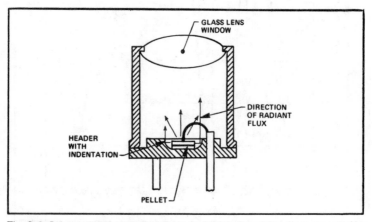

Fig. G-6. Schematic arrangement of a typical RCA IR-emitting diode mounted in an OP-17 package (scale exaggerated).

198

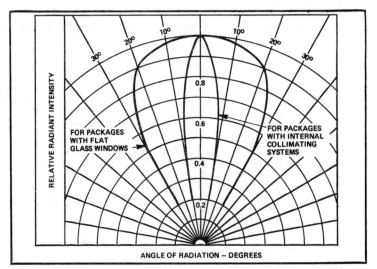

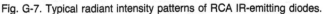

Fig. G-7. Typical radiant intensity patterns of RCA IR-emitting diodes.

to form partially reflecting optical surfaces. The adjacent sides are sawed to complete the rectangular structure and to suppress lateral modes. See Figure G-9.

The key parameters affecting the performance of a laser diode are the threshold, current density, J_{th}; the external quantum efficiency, η_{ext}; the emission wavelength; and the

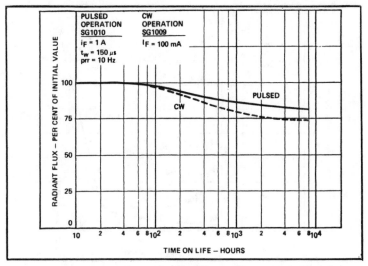

Fig. G-8. Typical life test data for RCA SG1009 and SG1010 IR-emitting diodes.

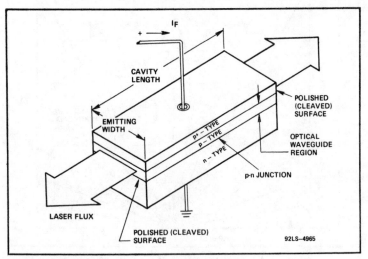

Fig. G-9. Typical injection laser diode structure.

diode radiation pattern. The threshold current density is the minimum current density through the device needed to obtain lasing at a given temperature. When the diode is driven beyond J_{th}, the radiant flux emitted is usually a linear function of the current and the slope of the output flux versus drive current is defined as the external quantum efficiency, η_{ext}.

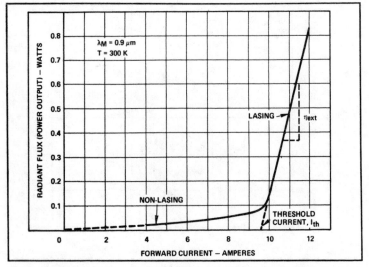

Fig. G-10. Radiant flux as a function of forward current for a typical injection laser diode.

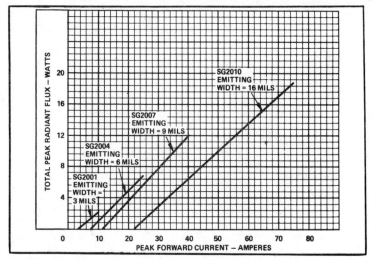

Fig. G-11. Typical peak radiant flux (power output) as a function of peak drive current.

Figure G-10 depicts the output characteristic of a typical injection laser diode.

RCA manufactures both single heterojunction laser diodes (GaAs types) and double heterojunction laser diodes (GaAlAs types). Their characteristics are discussed below.

Single Heterojunction Laser Diode. Gallium arsenide single heterojunction laser diodes are designed for applications where low duty cycle and high power output are required. They must be operated in the pulsed mode. The different types differ from one another primarily in the emitting junction width of the laser pellet. This width determines the laser drive current and peak output power. Figure G-11 shows typical peak radiant flux versus peak forward current for several RCA GaAs single heterojunction lasers. The curves are terminated at their maximum drive current ratings. The peak radiant flux of these diodes is generally limited to approximately one watt per mil of emitting facet at a 200 nanosecond pulse width to avoid catastrophic damage. Accordingly, a 9-mil diode such as the SG2007 is rated at a maximum of 10 watts peak output.

At the maximum duty factor of 0.1 per cent and the maximum pulse width of 200 nanoseconds, a repetition rate of 5 kHz is allowed for these devices. Because heat generation is

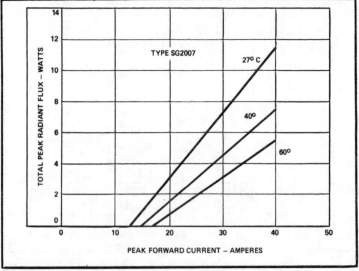

Fig. G-12. Typical peak radiant flux (power output) vs peak forward drive current showing the effect of temperature on threshold current.

the factor which limits diode operation, a direct trade-off can be made between pulse width and pulse repetition rate. At narrow pulse widths, in the order of nanoseconds, repetition rates of 100 kHz have been obtained with this type of laser diode.

Laser performance is dominated by the variation of threshold current with temperature. At temperatures above 27° C, threshold current increases (it about doubles at 65° C), and with constant drive current, the peak radiant flux falls to about 50 per cent of its room temperature value. Figure G-12 shows the effect of temperature on output of a typical SG2007. The structure of a typical laser diode is shown in Fig. G-13.

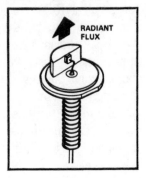

Fig. G-13. Schematic arrangement of a typical laser diode.

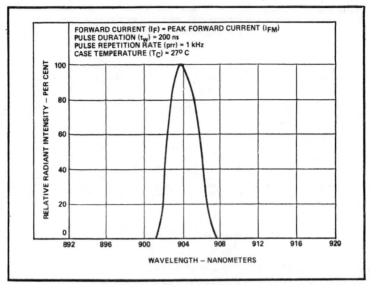

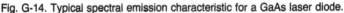

Fig. G-14. Typical spectral emission characteristic for a GaAs laser diode.

A typical spectral emission characteristic for a GaAs single heterojunction laser diode is shown in Fig. G-14 and the optical characteristics for this device are summarized in Fig. G-15.

Stacked diode lasers and linear series-connected laser arrays employing single heterojunction laser diodes are also available.

Stacked Diode Lasers. Two basic difficulties in achieving high peak power output from single laser diodes are (1) high drive currents are required to drive the larger laser pellets and (2) the larger source requires large, costly optics.

RCA's approach to the problem of achieving maximum optical power densities with practical drive conditions is the stacked laser device which consists of two or more laser pellets stacked one on top of another to form a compact emitting source. A typical structure is shown in Fig. G-16.

Laser Diode Arrays. RCA also mounts laser diodes in arrays to provide increased power outputs at reasonable drive currents. The individual diodes are series connected by small wires from the top of one diode to the back bar of the next; all of these connections are alloyed at one time. The laser array modules may then be connected in series parallel arrangements as shown in Fig. G-17.

Characteristics	Performance	Comments
Center Wavelength	904 nm	Temperature Sensitive 0.25 nm/° C
Spectral Width (50% point)	3.5 nm	Broad relative to other types of lasers
Half Angle Beam Spread (50% point)	9°	Total Collection requires F/1.0 optics
Source Size	0.08 mil x emitting width	Line Source

Fig. G-15. Typical optical characteristics of GaAs single heterojunction laser diodes.

In general, data accumulated on commercial GaAs single heterojunction laser diodes operated at current densities of 50,000 A/cm^2 and 0.7 to 1.0 W/mil of facet indicate that good long term operation is possible under conditions of typical laser usage. Figure G-18 shows typical life test data for commercial 6-mil laser diodes (SG2003 and SG2004) when operated at a forward current of 25 amperes (c.d., 53,000 A/cm^2, pulse widths of 200 nanoseconds, and a pulse repetition rate of 5 kHz). A decrease of 20 to 25% in peak power output is observed after 1000 hours of operation at maximum duty cycle.

Double Heterojunction Laser Diode. Gallium aluminum arsenide double heterojunction laser diodes are characterized by very low threshold current and may be operated in either the CW or high duty cycle mode at room

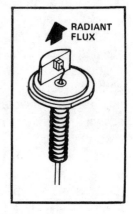

RADIANT FLUX

Fig. G-16. Schematic arrangement of a typical stacked diode laser.

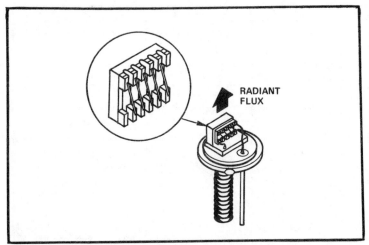

Fig. G-17. Schematic arrangement of a typical laser array.

temperature. Their wavelength peak emission of 820 nanometers is well matched with the spectral characteristics of most commercially available glass fibers as well as with silicon photodiodes. A typical spectral emission characteristic for this type of device is shown in Fig. G-19.

Double heterojunction laser diodes have inherently fast response time (less than one nanosecond rise time) and are

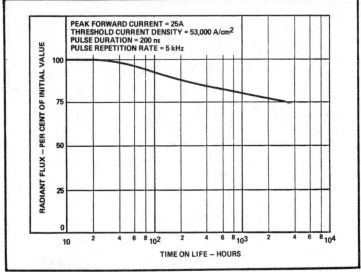

Fig. G-18. Typical life data for 6-mil single heterojunction laser diodes (types SG2003 and SG2004).

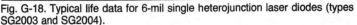

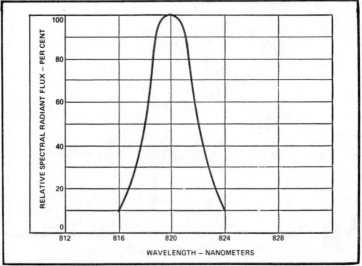

Fig. G-19. Typical spectral emission characteristic for an AlGaAs laser diode.

capable of digital and analog bandwidths in excess of 100 megahertz. These characteristics, as well as small source size—typically 0.0005″ by less than 0.0001″—, and high radiance make them particularly attractive for use in fiberoptic systems.

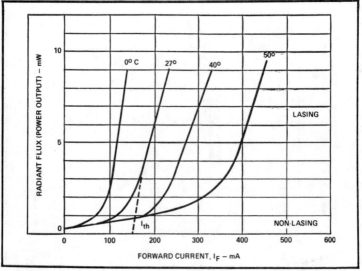

Fig. G-20. Typical radiant flux power output vs DC forward drive current showing the effect of temperature on threshold current.

206

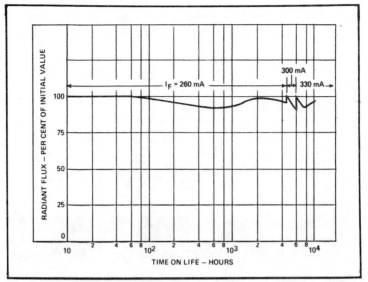

Fig. G-21. Typical life test data for a double-heterojunction AlGaAs laser diode type C30127.

Figure G-20 shows typical power output as a function of DC Forward Current for a typical GaAlAs laser diode. In most cases it is desirable to maintain the output of the laser at some constant level. However, as shown in Fig. G-20, a relatively small change in temperature can result in a significant change in device output. In some cases, the threshold current can increase to a level sufficient to cause non-lasing. Because of this strong dependence of threshold current on temperature for this type of device, the laser should be operated at some fixed temperature within its operating range. Proper operation can be obtained by mounting the laser diode on a heat sink whose temperature is regulated by a thermistor-controlled thermoelectric cooler.

The laser diode may be operated directly from a power supply. However, before such operation is effected, the power supply should be thoroughly checked for transients. Exposure of the diode to even very brief transient current spikes can cause catastrophic device failure. Safe operating considerations require that the device be protected by connecting a resistor (5 to 10 ohms) in series with the laser diode.

Typical life test data for a double heterojunction injection laser is shown in Fig. G-21.

Index